Planning and Creating Successful Engineered Designs

Planning and Creating Successful Engineered Designs

SYDNEY F. LOVE

President
Advanced Professional Development Incorporated
Los Angeles, California

ADVANCED PROFESSIONAL DEVELOPMENT, INC.

LOS ANGELES

Library of Congress Catalog Card Number: 83-71588
ISBN: 0-912907-00-2

Revised Edition, 1986

Manufactured in the United States of America

Published by:
Advanced Professional Development Incorporated
5519 Carpenter Avenue
North Hollywood, California 91607

Original hardcover edition was published by
Van Nostrand Reinhold Company

Library of Congress Cataloging in Publication Data

Love, Sydney F.
    Planning and creating successful engineered designs.

    Includes index.
    1. Engineering design. I. Title.
TA174.L68   620'.004' 2   83-71588
ISBN: 0-912907-00-2

Foreword

There is always a conflict between the needs for discipline and the needs for creativity in engineering design. This conflict can only be resolved by people who understand the engineering design process. Unfortunately, this process has not been studied and taught sufficiently in universities to produce a coherent and agreed upon system. But enough progress has been made by independent thinkers, such as the author of this book, that a great deal can be said on the subject and said usefully.

Engineering differs from science in that it always involves schedule and cost. It also always involves organization; and if the people in the organization are to pursue the same ends, they need to understand the process in which they are engaged. This book should go a long way toward helping the leader of an engineering organization to guarantee that all of the people in his organization understand the process. Syd Love has produced a book which covers the basic ideas of engineering design and discusses them in a fresh and lively style. I recommend it to anyone who wishes to become acquainted with the broad principles which underline the practices and processes of engineering design.

MYRON TRIBUS
Director
Center for Advanced Engineering Study
Massachusetts Institute of Technology

Preface

The theme of *Planning and Creating Successful Engineered Designs* is that engineering can be disciplined to provide engineered designs that meet human needs. During my first 20 years as an engineer, I was convinced (right from day one!) that I knew how to design. I knew how to engineer in the technical sense. Along the way, I picked up some engineering economics and a smattering of market savvy. But, as time went on, a few things bothered me, especially when trying to train subordinates. I was up against the same arrogance and ignorance that I had started with myself!

Fortunately, when I went back to university to get an advanced degree, I made contact with the new Department of Design at the University of Waterloo. A whole new world opened up. I discovered that the methodology of engineering design had become a discipline in itself. When I studied the new literature on systematic design methods, I liked it because it was congruent with my years of practical experience. It really made sense. Here was a body of literature that put it all together—but the practical world was not really aware of it. Later, when I set myself up as an independent consultant, I knew that my unique expertise was that I was one of the few experts in the design process who also had a substantial practical background. As you will see, this book bridges the gap between academia and the real world of engineering.

George Soulis, a professor of systems design at the University of Waterloo, has been my guru, mentor, and associate. He has that rare ability to ask questions that get me thinking on my own. When I first wrote a series of practical articles on the engineering design process, he read them and asked key questions. These early articles were clearly modeled on a book he coauthored, *The Discipline of Design*. That was 10 years ago. Since then I have gone my own way, improved the way I present the engineering process, and added my own original ideas on needs analysis, trade-offs, decision analysis, optimization, and communication.

All of the material in this book has been subject to the test of reduction to practice. It worked for me. In addition, it has been tested against the seasoned experience of hundreds of senior engineers and engineering managers who attended my seminars and demanded clarification when it was needed.

As I search my own experience for practical examples to make a point, one other person keeps coming into the picture. That person is John Sennik. He was a practical engineer who went to the school of hard knocks. He did not have any degrees,, but he had that rare ability to generalize on experience. He had an intuitive understanding of the engineering process. Because he worked with me, he undoubtedly lives on in many of the points made in this book.

ADDENDUM FOR THE REVISED ADDITION

Computer are enhancing our capacity to design. The advent of Computer Aided Design (CAD) has added the power of visual graphics to the imagination of the mind. Hooked up together, the designer can do many jobs better and faster. The power of instant iteration is making the difference.

At the time of writing, most engineers, scientists and technical personnel have access to both large computers and personal computers. In a few years, the use of the computer to aid synthesis and analysis will be the norm. Marvelous software packages are being developed to aid every facet of design.

The people preparing engineered designs of all kinds will be aided by computers. These same people will see new computer applications and have software developed to aid them even more.

It is with this in mind that I have added a chapter on Computer Aided Design. To do this, I was fortunate to be able to get Dave Hogg to write this chapter. His full-time job is to assist enterprises to add computer power to their designing, drafting and manufacturing. He is the Manager of Educational Programs and services for the Ontario CAD/CAM Centre in Cambridge, Ontario, Canada. His experience in this area towers above that of most of us. From him we can learn.

Sydney F. Love.

Contents

Planning and Creating Successful Engineered Designs

1 Making the Design Process Work For You

KEY IDEA: AN OUNCE OF DESIGN REVIEW AT THE BEGINNING IS WORTH A POUND AT THE END.

A PERFECT DESIGN IS A MYTH

I well remember the first time I learned about the importance of iteration. I had just joined a firm which had put me in charge of the Television Engineering Department. Over the next few weeks their troubles unfolded before me and I was expected to apply some mysterious healing salve that was going to make everything right.

A previous television design had given them much field trouble, and now the whole of the finished stock was going through production again. This time they were reworking them to a modified design which was supposed to end all their troubles.

I asked for test results as proof of the modified design, but there were none—only blind hope. As it turned out, we never ceased to be hounded for field improvements in that design. In the light of history, we would have done well to scrap the whole design and all the TV sets produced therefrom.

All the same, it did teach us something about design—*never rush into production with a design that has not been tested out beforehand, because design iterations in the production phase or use phase are very expensive and very constrained.*

1

My experience in the radio tube industry paid off in this new job. Before going into full-scale production on a radio tube, the engineers always ran a pilot run under production conditions. From this they got proof that the design was reasonably correct (or not). If it was a revolutionary design, they ran preliminary laboratory type pilot runs where the new design was made up by hand. This was the iteration principle that I applied to our new television designs.

After a few years we worked out a good plan of formal iteration which generally resulted in successful production runs. Our first design phase was on paper with circuit diagrams. Calculations were performed. Then a "breadboard" of the design was put together on a wood and metal chassis with hookup wire holding it together. This tested out the basic design concepts.

After the bugs were shaken out of this breadboard, we would then go to a handmade prototype which was very close to a production unit. Before the first prototype was completed, another one would be under way so that we had a standby unit for life testing to check out the reliability of the design. When the two prototypes were completed and tested, and the engineering drawings issued for final production, we then constructed six units of the final design drawings.

These were put together by experienced production workers. Everyone's advice was taken seriously and all necessary improvements were then put into the design in time to be incorporated in the following pilot run of 100 units. The 100 units were then field tested thoroughly and any further revisions found to be warranted were then put into the design just before full-scale production.

It would be wrong to assume that full-scale production was without some problems. Maybe the production manager was unsophisticated enough to think the design would be perfect and without some problems, but we designers always knew there would be some unforeseen problems. However, by our process of formal iteration, by revision and review, we had shaken the major bugs out of the design before full-scale production took place. The remaining minor iterations in design were usually a result of information being discovered during the production process—faults which had not surfaced during the actions taken beforehand. However, after a few production runs, we could say that we had a mature design and it was only necessary for production to reorder the parts and repeat the design as it was.

The only factor then making us change a design was the obsolescence factor. In other words, the design might be technically perfect but still not satisfy the changing needs of the user. So we would go through a couple of years of updating the existing design to keep it competitive—our final iterations.

WHAT YOU WILL GAIN FROM THIS CHAPTER

In the following development of the *Principle of Iteration* you will see that what is sound practice in designing is based on a fundamental principle of the design process. Iteration is very important in all phases of the design process, whether it be for a mass-produced product, or for a one-of-a-kind, large-scale project such as a hydro-electric dam.

Objectives

    1. You will learn to use a fundamental principle of design: the Principle of Iteration.
    2. You will learn the value of cybernetic feedback in the design process.
    3. You will learn the need to have a plan of formal iterations.

Benefits

    1. You will improve your success rate by reducing the probability of an expensive redesign or rework.
    2. You will know what to do to prevent a design disaster.
    3. You will have a sound rationale for countering those organizational forces which try to keep you from iterating when you should.

In this chapter, you will examine some self-evident truths about design and generalize with me up to the useful Principle of Iteration. This principle applies at all times when you design. You will be shown examples of how planned iteration can be applied to a technical equipment design, a mass-produced product, and a large-scale engineered system. These will help you apply the principle in your own area of endeavor. Then follows a case history of a product design that was aborted before it became a disaster.

The essence of this chapter is neatly summed up with specific do's and don't's. This will introduce you to Design Guidelines which will continue in every chapter of this book. After reading and studying the book, you will need only to refer to the guidelines because you will understand them and believe in them.

At the end of this chapter are exercises to help you consolidate your knowledge and become proficient in using the concepts of this chapter.

This chapter covers the way that material resources are harnessed to meet human needs. The seven systematic steps near the end of this chapter are to be elaborated upon individually in the following chapters of this book.

THE PRINCIPLE OF ITERATION

Throughout this book, we will use the method of science to arrive at principles and guidelines. We will look at a number of real world situations which are related, and then generalize on them. This is the process of induction. From that we will come up with a principle which is *a condition plus a generalization.* After the principle has been developed, we will use deduction to derive applications and specific guidelines for action. Now, let us do it for this chapter.

Informal Iteration

Designing something new is like being on a voyage of discovery. As you progress into the design you discover more and more that you did not know in the beginning. You discover that things do not work so well when put close together. You may need to go back to plan again, thereby doing an informal iteration. To go back to any previous step in the process is to iterate. In designing, it is the name of the game. If you do not iterate, you will not make use of new information, and you will not be able to come up with something better than before. It is through iteration that designs are made good.

Some of my youthful attempts at constructing radios were quite exciting to me. I would get circuit diagrams from friends or from a book, gather the parts together, then wire up the radio and hope that it would work. What great elation I had when the radio actually played! On closer inspection, I would find strange whistles at places on the dial where they should not have been, and occasionally it would sound like a motorboat. I very quickly learned that certain parts should not be placed close together because of electromagnetic coupling.

In later life, as a manager of a group of engineers responsible for designing TV sets, I found that nobody, but nobody, could go from a conceptually new circuit diagram to a completed chassis without some design difficulty. Electronics is the assembly of components to a 3-D array which gives rise to all kinds of noises and squawks which cannot be predicted by circuit diagrams or by circuit analysis. In other words, our mathematical models are inadequate to predict all that will happen when we bring an idea to reality. It is only by bringing parts together that we discover things "we should have known all along," but in fact did not.

Take, for example, a matter of designing an automobile engine. Very careful analysis will home-in on approximate sizes of components. Then, if followed up by careful spatial design on a drawing board the parts will probably fit together.

However, strange things happen in the testing of an automobile engine. The heat from one part affects another. The vibration of one part affects another. There is an interactive term between components which is unpredictable. This makes it difficult to predict the optimum size of a bearing. Sure, anybody can specify an oversize bearing, and if cost is no problem, it will work forever. But to make an economical engine that people will buy is not to over-design. There must be a balance so that all bearings will have more or less the same safety factor. Invariably, the life testing of machines will show that some bearings will need to be beefed up and that others can be reduced a bit. This "balancing" of the mechanical design is part of the "art" of design which we so frequently speak of.

If we look at the history of design disasters we will often find that the engineers were not allowed to iterate. They were forced to march in a straight-line fashion towards a certain completion date.

The design was ready and the parts fitted together, but unforeseen problems developed during assembly, testing and using. Where it might have taken a few days and a few bucks to do an iteration early in the design process, the cost of making the same iteration downstream had escalated enormously.

The point you should remember is the following:

ITERATION IS NORMAL AND DESIRABLE.

Formal Iteration

When I have presided over post-mortems of design disasters, I have frequently found that the "unknown" that gave us trouble was known way back in the design process. However, the engineer did not, or was not able to, act on the information; or he just took a calculated risk—to everyone's regret. The design changes that should have been made were set aside.

I would go as far as to say that many a time when I personally have had a design difficulty, it was really a rediscovery of what I had uncovered earlier in the design process. My intentions were good, but my memory was less than perfect—and the same goes for you!

Large corporations have discovered in their design post-mortems that there was in fact information known beforehand that should have been used. So they decide, "never again," and then they schedule formal review meetings for every major design. At review meetings, all aspects of design are looked at, test results are scanned very carefully and suspected problems are dealt with. Most large corporations have formal procedures which spell out the frequency and timing of formal review and revision meetings.

At these formal iterations they decide whether or not the design will go ahead, be revised, undergo major changes of direction or be aborted. Design targets often need to be changed because the users change their minds about what they want. Some design solutions take several years to bring about and the needs actually can change.

Formal iteration is an organization's way of insuring that informal iteration is not overlooked and that the supposed needs be reviewed. Table 1.1 is a comparison of those names given to phases in various situations, going from a simple lab instrument to a complex military procurement.

You too can learn the design tip that big organizations have learned. Plan to have iterations. Use meaningful points in the design process. Formal iteration does not mean having a formal review every six months, or every month or at some periodic interval. The formal reviews should fall into natural nodal points of the design process; that is, at the ends of some particular phases where a lot of

TABLE 1.1: **Formal Phases in the Design Process.**

Mass Production	**Construction**
Concept	Proposal
Feasibility	Site plan
Preliminary design	Detailed design
Detailed design	Construction
Pilot Run	Remodeling
Use	

Military System Procurement	**Book Development**
Concept	Idea
Feasibility	Synopsis
Configuration	Chapter headings
Detailed design	Chapter outlines
Production	Detailed chapters
Deployment	Typesetting and proofreading
Retirement	Publication
	Revision

Dr. Glockenspiel's Design Phases

Euphoria
Disenchantment
Search for the guilty
Punishment of the innocent
Distinction for the uninvolved

information is to be available. If the test results are delayed by two months then the formal review is delayed by two months.

You can also learn from the mistakes of big organizations in their use of formal iterations on every design. The number of formal iterations is dependent on what is being designed and should be developed for every unique situation.

THE PRINCIPLE OF ITERATION.

When new factors are involved in an engineered design, then iteration is both normal and desirable throughout the design process.

THE COSTS AND RISKS OF PHASES

Figure 1.1 is a graphic representation of formal iterations in the process of design. We will call the early phases the "upstream" phases of the design process and the later ones will be called the "downstream" phases.

Upstream, the design phases don't cost very much. They can be in the mind, or on paper, or on cheap scaled-down models. We are generally innovating while upstream, and the probability of any new design coming downstream is less than 100 percent. There is always some risk of failure. There is also a risk that the projected design will not be needed when ready.

As we proceed downstream in the design process, we find that the cost per phase goes up very much, but the risk of not going on to another phase is greatly reduced. One highly innovative company reported that of 1000 proposed products, only 100 were designed, of which 10 were produced and only one was a really successful new product. This is a high-risk business!

At the other extreme, in low-risk designing, like public utilities, the success of each design phase is so certain that the risk of not going on to the next phase is not even considered.

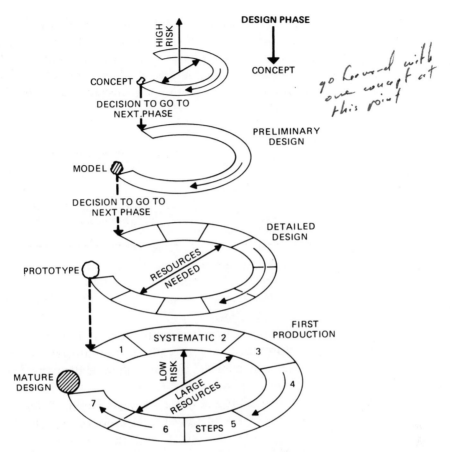

Figure 1.1. Formal Iterations in the Design Process.

I have argued that iteration is vital in design. How many formal iterations should you have? That depends on the nature of the design. How do you make use of the principle of iteration? The answer is to *iterate as much as you can upstream and as little as you can downstream*—one will help the other. It may seem to delay the completion but if you discover information early in the design process, you will not really be delaying the overall time for a fully satisfactory design solution.

UPSTREAM AND DOWNSTREAM DESIGN

A design project in its early phases is an upstream project. Concept and feasibility phases are characterized by a high risk of not going on to the next phase. The project may never be successful enough to reach a downstream phase. Moreover, the initial direction of a project may change substantially as new information develops during the design process.

A problem we have with upstream projects is that there is much uncertainty about what can be done in any given unit of time. This is especially true in R&D. Perhaps this is why in R&D we seem to be opposed to any form of planning. We say that we cannot accurately predict when a given bit of applied research can be completed or when a new idea can be made to work. Forget it! This is muddled old-fashioned thinking and we must break away from this habit. The successful project of putting a man on the moon (safely, on time, and within budget) is ample evidence that planning can be done for upstream activities.

Here is another difference from downstream projects. While upstream, the performance targets can be recognizably higher than those we are likely to achieve. We are striving for technical excellence. Later on we may relax the target specs. On the other hand, when downstream in a project the performance targets become "set in concrete" so to speak and they resemble the final specifications. It is barely possible to make technical tradeoffs downstream in the design process. For this reason we should be sure that tradeoffs are properly guided during the upstream phases when tradeoffs can and will be made. This will be covered in more detail in chapter five.

THE WORDS FOR THE PHASES MAY VARY
BUT THE PRINCIPLE IS THE SAME

Each organization has its own household words for the formal design phases. Some of these are defined by their internal procedures, others are informally understood throughout the discipline. That is why I do not give you "official" names to use for design phases. Table 1.1 gives you examples only. It is better that you use the words which are understandable to the organization with which you work. Just make sure you understand the principle.

APPLICATION TO A TECHNICAL EQUIPMENT DESIGN

An Electrical Substation Transformer

PHASE	COST	RISK (OF NOT GOING TO THE NEXT PHASE)
Define Requirements		
Customer and supplier jointly develop specs.	*$500*	*1 in 100*
Detailed Design		
Engineers and design draftsmen work from historical data they accumulated on other designs. No prototype.	*$5000*	*1 in 10,000*
Production		
Construction in a workshop under supervision of the design engineer and production personnel.	*$50,000*	*1 in 1,000,000*
Installation		
A contracted crew transport, install and connect.	*$2000*	*1 in 10^9*
Use	*$1000*	

APPLICATION TO A MASS-PRODUCED CONSUMER PRODUCT

An Electric Toothbrush

PHASE	COST	RISK
Concept		
A product planner draws up a proposal	*$100*	*1 in 10*
Feasibility		
Product planner and design team assess cost and market	*$500*	*1 in 10*
Preliminary Design		
Engineering and Industrial Design build a prototype for testing	*$2000*	*1 in 10*
Detailed Design		
Final prototypes and drawings are produced	*$10,000*	*1 in 100*
Pilot Run		
100 units are produced using production technique	*$10,000*	*1 in 1000*
Production		
100,000 units are produced for sale	*$1,000,000*	*1 in 10^6*
Use		
Final modifications are made to improve producibility and lower the unit cost	*$10,000*	—

APPLICATION TO A LARGE-SCALE CAPITAL
CAPITAL PROJECT

A Hydroelectric Dam

PHASE	COST	RISK
Feasibility *Sites are studied and the needs defined.*	*$100,000*	*1 in 100*
Preliminary Design *Sites are analyzed and final costs estimated.*	*$250,000*	*1 in 10*
Detailed Design *Complete sets of specifications and drawings are prepared for tender.*	*$5,000,000*	*1 in 1000*
Construction *Materials and equipment are purchased and the dam is put into operation.*	*$50,000,000*	*1 in 10^9*
Use *Maintenance and minor modifications.*	*$5,000,000*	—

A CASE HISTORY—A DESIGN
THAT DIDN'T WANT TO BE

This case history is one of a large and successful product manufacturer. It is an example of reasonably good planning and sensible
use of the design process.

Concept Phase

This phase took place in the mind of the marketing manager who recognized the lack in his product line of an 11-inch portable TV. His major competitor had models which used new compact devices that enabled a very light-weight TV to be made. In his concept phase, he may not have explored many alternatives, but his thinking did result in a decision to write a document called a "New Product Opportunity." Time: about one day. Cost: about $100.

Feasibility Phase

A product planner surveyed the market potential and competitive efforts. Using estimated tooling and engineering costs, he analyzed the potential profit. He outlined the performance, weight and target costs in a quantitative way—the development of target specifications. Engineering then looked at alternative design possibilities, having first evaluated competitive models to set target specifications for sensitivity, brightness, contrast, etc.

The styling manager came up with sketches for eight styling concepts which were reviewed for feasibility by engineering and marketing. The financial feasibility was determined by the relationship of the tooling costs to the configuration of the tuner, speaker and picture tube. At a review meeting the performance and cost objectives were revised and refined. New time objectives were established. The feasibility phase ended in a document called the "New Product Specification." Time: about one month. Cost: about $2000.

Preliminary Design

The design engineers now took over, and one of them worked on the problem of using a power transformer, or a series connection, for the tubes. A technician developed a new wiring style which reduced the cost and weight at the same time. A prototype was constructed and demonstrated to the marketing manager. Because of the use of a small transformer, the weight was worse than the initial target and the cost was marginally higher.

The marketing manager, however, thought that the feature of

having a power transformer offset the cost and weight disadvantages, and approved of this alternative design proposal after consulting with some of his key customers. Time: about 3 months. Cost: about $10,000.

Detailed Design

The engineers started on the preparation of detailed mechanical and electrical drawings. However, by then various members of the design team sensed that the market was turning sour. The introduction of new light-weight transistorized TV's by foreign competition would make this new 11-inch tube-type TV set obsolete shortly after its introduction onto the market. This intelligence about the new technology was communicated to the marketing manager who resurveyed the market and agreed that the situation had indeed changed.

The whole project was aborted. Since it had ended at the initiative of the design team, they were not upset. On the other hand, even if there had been some disappointment with the ending of the project, a greater disappointment in the future, at a much higher cost, was avoided by aborting the project at an early phase. Projected cost of this phase was to be $50,000 over a six-month period. The actual cost was $20,000 over the three-months up to the abort point.

SYSTEMATIC STEPS WITHIN A DESIGN PHASE

To properly complete one formal phase in the design process, you should consider each of the following steps in a systematic way. Each should be considered, although it is not always appropriate for each to be done in detail. It is not even necessary to do them in the order presented, but it is advisable that they all be considered prior to the termination of one of the phases of design. These steps are as follows:

1. Analyze needs
2. Set objectives

3. Develop target specifications
4. Create alternatives
5. Screen for feasibility
6. Select the solution
7. Communicate the design solution

All of these design steps will be treated in detail in the following chapters. You will learn to apply them like a pro.

A professional will plan out the design process with an adequate number of upstream phases. Within each phase he will systematically check all of the seven design steps. The pro knows that it is a good idea to enter each phase by rechecking what the user wants. This means that he reconsiders the step of "analyze needs." If the user's needs have changed or even if his perception of them has sharpened, then the objectives themselves may also be changed. All seven steps are reviewed.

CYBERNETIC FEEDBACK IN THE DESIGN PROCESS

Designing can be treated as a process where the input is a problem and the output is the solution, as shown in Figure 1.2.

The design steps outlined above are a process which generates a solution with respect to a problem. According to the logic of these

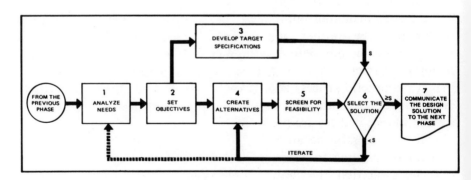

Figure 1.2. Seven Systematic Steps Within a Design Phase.

steps, the "Develop Target Specs" (also called "Develop Design Criteria") is not actually needed until just before step 6 where they are used in the selection of the solution.

Nevertheless, if you wait until you have a solution before drafting up a set of target specs it is quite likely that your specifications will fit the solution you have in hand better than they fit the problems with which you started. There is a tendency to become enamoured with one's own creative solution and to assume that this is what the users want. For this reason, the best position for developing target specs is close to the problem end of the process; that is, immediately after the "Analyze Needs" and "Set Objectives." The specifications so developed are likely to fit the needs, whereas if developed after an apparent design solution is at hand, they may really be nothing more than the measurements on a design prototype. (This error frequently happens in the design of mass-produced products, but you can see that it would seldom happen if the design was contracted out to another organization.)

Here is another advantage. If your target specs are representative of an ideal solution to the problem, you will force yourself to keep looking for new alternatives until you come up with one which fits the target specifications—and not make any compromises that haven't been thoroughly checked out against the user's needs.

Take a look again at Figure 1.2. You will see that the target specs which are developed near the beginning are held until the alternatives are being checked out. If no such solution is found, you must iterate back in the search for alternatives. This disciplines you to be more creative. In the long run, if further alternatives are not producing a substantial improvement, you can still go back to an earlier step and take another look at the target specs. Maybe you are a bit too idealistic, but before changing the specs you should check the analysis of needs.

A cybernetic process adapts to the environment. In this procedure the target specs are a model of what is required in the environment (the users). If these are properly met, then the designing organization will be successful because they are providing that which the environment is prepared to exchange resources for. If, on the other hand, it is believed that the desired resources would not materialize, the cybernetic designer is usually able to try other solutions to adapt to the environment. The feedback time constant for product design is rapid because of profit measurement. For civil engineering works produced by the government, the time constant is decades because of the intervening political process.

FOR SUCCESS IN ENGINEERED DESIGNS— DO THESE THINGS

I have shown you that iteration is normal and desirable. Because it is desirable, you should make it happen in the right places. The quintessence of what has been said in this chapter can be summarized into a few guidelines.

DESIGN GUIDELINES

GUIDELINES FOR THE DESIGN PROCESS.

Do

Do plan to have low-cost iterations upstream in order to avoid high-cost unplanned iterations downstream.

Do encourage your colleagues and subordinates to start design iterations when they should.

Do develop your target specifications before searching for alternative design solutions.

Don't

Don't rush pell-mell into the design process and try for a quick shot at the target. Heroes are not made by design disasters.

Don't scream too loudly when somebody rips up a design and starts all over. It won't hurt nearly as much as doing it later on.

Don't wait till the last minute to issue your specifications. The designed object and the specs can be compatible, but your nemesis may be just down the road when they find out that you solved the wrong problem.

WHERE TO GO FROM HERE

Okay. Now you have my design guidelines and you know what to do. Right? Not quite! When do you think these ideas will be yours to own and use to your advantage? Only when you apply them to your own work! In a few weeks, you will probably remember only 20 percent of what you have read. With repeated reviews you can get that up to 40 percent. These ideas will only be your tools when you finally apply them to your own designs. Then you will have 100 percent retention.

If you really want to get ahead with engineered designs and you want to be a pro or super-pro, then put these ideas immediately to work on your own design. If you cannot do that conveniently, then take some time to practice on the exercises which follow.

FOLLOW-UP EXERCISES

Questions to Help You Consolidate Your Understanding of this Chapter

1. What is the difference between formal iteration and informal iteration? Can the latter be scheduled?

2. If you plan out a design project with a series of formal iterations or phases, is that all the iteration you can expect?

3. Within a design phase there are seven steps. These should be repeated to some degree in every design phase. Why is it advisable to repeat the step called "Analyze Needs?"

4. The step called "Develop Target Specifications" follows the "Set Objectives" and precedes "Create Alternatives." Logically speaking, one does not need the specifications until step 6, "Select a Solution." What is the reason for developing the target specifications prior to creating the alternatives?

Questions on the Case Study

On a piece of paper, make yourself a sketch of the open discs shown in Figure 1.1. Leave off the wording. Now study the case history of the TV set design that was aborted and label the phases and the item produced at the end of each phase. Indicate the time and cost of each phase and the point where the project was aborted. Study this and record what you would recommend as a procedure which may have led to the project being cancelled earlier, thereby saving on some design expense. (Hint: See if any of the Design Guidelines apply in this case.)

Assignments To Help You Become a Pro

1. In my seminars and courses on the use of the design process, I have used a "practice project" for persons to apply the concepts. It is usually something between my examples and what they do in their own work. This would be a great way to consolidate what you have learned. On the other hand, if you have a real design project that you can carry through from A to Z, then that is even better. Let us get on with selecting a practice design project from the world of possibilities, including any you are working on right now. Here are the guidelines for selecting a practice project:

   a. A good practice project is one which does not get you hung up on understanding the technology. Technical problems tend to direct your attention away from the basic design methodology which you are trying to learn. Choose a project with a simple technology that you understand, or alternatively, with a technology so advanced that you can not get involved with technical details. If you are doing this with a group of other persons, your design project should not belong to a group member, or else that person will tend to dominate the proceedings because of the technology involved.
   b. If this is a practice project, do one on something you know enough about without calling in technical consultants. You should be able to make enough assumptions to do the work of each chapter in about an hour's time and frequently less than that.
   c. Preferably, choose a design topic which will enable you to visualize all the phases in planning and enable you to complete a concept phase while you are studying this book.

d. Examples of projects which have worked well for groups are as follows: garbage collection machines; safety devices; water safety equipment; remote control lawnmowing machines; manned space-lab in stationary orbit; etc. This list could be endless but you get the idea. You can choose a fun project.

2. Whatever your design project is, plan out all of the formal iterations that you think you should have. You can use Figure 1.1 as a guide, or do it in tabular form. You should end up with the number and names of the phases, the approximate cost of each phase, the time from beginning to end of each phase, and the approximate risk of not going to the next phase. You can use one of the applications as a model for your exercises.

3. If you are practicing engineering design, you probably have been responsible for, or have been associated with, some project that did not work out right. Maybe it required a lot of redesign or a lot of rework, or the cost and time overruns were excessive—or it might have been a major disaster. Some of these things happened due to outside circumstances and there was nothing you could have done to prevent them. On the other hand, some were under your control. Take a good look at one or more of your case histories and see whether a little extra iteration upstream by modeling, simulation or whatever, would have increased the chances of success. If you have a good case history, then you have a very good example when you want to talk to management about needing more time and resources while still upstream on a project. You can then turn an unfortunate experience into something useful for all concerned.

2 Finding Out What the User Really Needs

RIDING HIS HOBBY HORSE

The executive engineer of a Toronto firm attended one of my seminar-workshops on engineering design. He worked for a firm that developed and manufactured very specialized geological survey equipment. Because few people made such equipment, it tended to sell itself. This gave them a cash flow that kept them in business.

The executive engineer could see what was technically possible and was working on a 20-fold increase in sensitivity. Although the marketing manager complained that his present line of instruments was becoming obsolete, the executive engineer kept the new products in development and would not release them to production.

As a result of the seminar, the executive engineer realized that to stay in business he must meet interim users' needs and do it right away! He told me that he had been riding his own hobby horse of the technical sophistication because this would enhance his reputation as a top-notch engineer. He had been resisting marketing efforts to put an interim product on the market because in a few more years he thought they would have all the answers.

After changing his viewpoint, he became more market-oriented. He reasoned that if the customers wanted an interim product badly enough, with only a five-fold increase in sensitivity, then he would give it to them. It would interfere with his tests and experiments on the advanced development but an interim product would be better for both company and customers in the long run.

Six months later he became the president of the firm. Being then responsible for marketing as well as for engineering and production, he made sure that his engineers were oriented toward fulfilling user needs rather than toward fulfilling themselves by technical sophistication.

A NEAT IDEA THAT SURPRISED
THE MANUFACTURER

An electronics engineer experimented in his basement and developed a neat feature for a television set. It was this: a microphone-speaker was wired from the baby's room to the TV set. When the baby cried, the program sound was interrupted and the baby could be heard loud and clear. If not inclined to dash to the baby's room, a concerned parent could push a switch and say comforting things to the baby—and still keep track of the TV program.

The electronics engineer obtained a patent on a low-cost electronic circuit that made this possible. Then he peddled it to TV manufacturers.

One delighted TV manufacturer bought the exclusive rights for a lump sum of $20,000. They designed a plug-in accessory to be sold and used exclusively with their TV sets. Technical tests showed that it worked beautifully. Enthusiastically they produced 2000 of the accessories. They also mounted an expensive promotion program to stimulate the sales of the TV set with the exclusive "baby minder" accessory.

One year later they reviewed the idea. They had sold four! The item was discontinued and the stock scrapped. What they had was a great idea, but nobody wanted it!

WHAT YOU WILL GAIN FROM THIS CHAPTER?

Objectives

1. *You should learn how to use a needs analysis matrix and do an imaginative search for potential user needs in relation to a proposed design.*
2. *You should learn how to judge the relative importance of various user needs.*

Benefits

1. *You will be able to oreint your design efforts to what users will, or would, pay for. You will, therefore, avoid the unhappy situation of having designed something that is really not needed.*
2. *You will be able to recognize when an adequate needs analysis has not been done, and thus initiate or call attention to the analysis that should be done.*
3. *You will be able to make rational decisions on which users' needs should be met by the design.*

In this chapter you will learn why designers win by meeting user needs. Other names given to this step are needs analysis, market study, or requirements definition. As explained in the previous chapter, this is the first systematic step within a phase. In the next chapter you will be shown how to convert user needs into design objectives.

By generalization of familiar examples, you will understand what the Principle of User Needs is all about. You will be introduced to a systematic way to find needs with the Matrix Analysis of User Needs. This guides you into every nook and cranny of what people may want in relation to a proposed design. Following this are practical applications to an electromechanical product, a made-to-order equipment and a large-scale capital project.

The scope of application is widened in the section which discusses some real case histories. Thus, you will most likely find an example akin to your own field of endeavor. Explicit directions are to be found in Design Guidelines for Needs Analysis, where you will learn what to do and what not to do.

In the follow-up section are questions to help you consolidate your knowledge of the Principle of User Needs. There are also exercises for practical application to projects of your own choosing.

DESIGNERS WIN BY MEETING USER NEEDS

The reason for making a design that suits the needs of its users is simple. Users have some control of the resources which will be exchanged for the product or service which is designed for them.

Ultimately this reaches to you and someone decides who designs what.

The simplest case of the exchange of resources is that of a product which is made and sold to the general public. Resources are exchanged for the product and the profit goes to the organization which developed the product. This encourages the investment of more money into the designing activity.

A similar situation exists for one-off equipment designs, although the time constant might be quite long because some heavy equipment, such as a special tree harvesting machine, may take years to develop and evaluate. Many other factors may intervene. The relationship between the success of the design and the users' needs may be a bit foggy, but in the long run the relationship still determines who will continue to design these machines.

Even in the public service this is true. Because of the democratic political process, government persons are generally responsive to the needs of the people when they use public money. Cost-benefit analysis is frequently employed. True, the time constant might be 10 or 20 years, especially for large-scale public works such as roadways and irrigation dams, but history is full of examples wherein a government finally did something about a department that was doing lousy design.

Some designers find that they are working for people who dictate what they must do. This is the most common complaint that I get from engineers when I talk about responding to users' needs. Still, after some discussion we usually agree that it will be best for all parties in the long run if the designer does his best to meet users' needs, whatever the circumstances of his work.

DECIDING WHICH NEEDS TO MEET

Suppose you believe that it is to your advantage to meet user needs. You soon find out that users have more needs that can be possibly met by any design.

What do people need in a car? They like a big trunk, plenty of seating space, super power and low up-keep costs. Most like power steering, power brakes, power seats, power windows, power tuning, and so on—all of these for a bargain price. You simply cannot design and sell a car that has everything that people want, or need.

To be practical about user needs we must realize that the users want some things more than others. "Utility" is the name given for value to the user of a design feature. In general, a car has high utility as a status symbol—styling, chrome, plush interior and gadgetry are geared to this symbolic need. People also have utility for power, economy, space etc., and for utility they are willing to exchange money.

In the case of public works, where people do not have a specific choice of exchanging money for what is designed, they still have preferences as to where their tax dollar goes. The government designers that win try to find out the preferences of the taxpayers.

Because the designer cannot possibly meet all users' needs then, a rationale is needed for deciding which to meet and which not to meet. Obviously, the best rationale is to satisfy user needs in proportion to the resources they would exchange for the various utilities that they get from the design. If they will pay for it, give it to them.

We are now ready for a general statement of the principle embodied in the above.

THE PRINCIPLE OF USER NEEDS

Since all user needs cannot be met by a design, those that are to be met should be determined by the resources that would be exchanged for the completed package of Performance, Time, and Cost.

HOW TO ANALYZE USER NEEDS

The three major dimensions of P, T, C: Whenever you start an analysis of user needs, it is convenient and advantageous to make your study in the dimensions which we will call Performance, Time, and Cost.

The performance, P, contains all those technical and nontechnical attributes of the design object itself. Under this dimension you will look for answers to the question: What should the design do when it is completed and in operation?

The time dimension, T, pertains to all time aspects of the design, such as: When is it needed? When will it be ready? When will it be worn out or obsolete?

The cost, C, pertains to all those money aspects of a design, such as the initial resources required to "get the show on the road"; the cost of labor, materials and overhead for actually building the design; the cost of transportation to the site and deploying it for use; the profit aspects if any; the maintenance cost, and so forth.

This leaves us with the dynamics of performance and of cost. We shall call them $\Delta P/\Delta T$ and $\Delta C/\Delta T$. During the life of a design, from conception to retirement, there are often requirements for changing the performance. There may also be requirements that the cost dimension change. For example, a product which is innovative has a higher initial cost of manufacture and consequently higher selling price than it will have when it is commonplace.

We shall use P, T, and C as one side of the matrix analysis of needs.

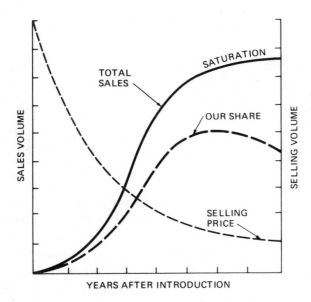

Figure 2.1. The Expected Market for a New Product Shows That a Reducing Selling Price With Time ($\Delta C/\Delta T < 0$) is a Need.

The other side of the matrix will be derived from the basic needs of people.

Basic Human Needs

All technical requirements of a design can ultimately be traced to a basic human need. For example, a generator might be specified with an output of 100 megawatts of power. This electric power in turn fulfills human needs for the preparation of food to satisfy basic physiological requirements; for the lighting of streets so that they can feel safe; for the powering of machines in man's service; for electric power for the TV which satisfies some psychological needs; or living room lights the enable people to sit around and talk to satisfy their social needs.

At one end of the design spectrum are highly technical designs where the technical requirements are known and there is no need to analyze the basic human needs. A conveyor might be one. It is required to move so many tons an hour from A to B. The designer is then faced with determining only the technical requirements, and probably will not need to examine basic human needs except on a special occasion where esthetics, noise, pollution or convenience are factors.

At the other end of the design spectrum are consumer products which interact very much with people. The car, for example, is designed to meet a basic need for transporation, but because it interacts with our lives so much, around 25 percent of what we pay is for the satisfaction of other needs. For any new product there is no existing book or manual that you can consult to find out what people need. An original study must be made.

While we have pretty good models of the physical dimensions of man, we engineers do not have suitable models for analyzing his other needs in relation to a consumer product. A model for human needs from management theory that I have found most effective in analyzing user needs is the "Maslow Model of Needs Hierarchy." This has been tested and used in industry and by managers so many times that we can say that it is almost universally accpeted in the management field. The Maslow Model as adapted to design does need the hierarchy but makes use of the categories shown in the matrix of Figure 2.2.

REAL NEEDS AND SYMBOLIC NEEDS

People have real needs for food, shelter and safety, and they also have symbolic needs associated with them. Breakfast foods are an example. Besides nourishing you, they symbolically make you vigorous, virile, and they have the most delicious crunch you ever experienced. Words we use for designs that meet symbolic needs are health symbols, status symbols, sex symbols, prestige items, and so forth.

Because a symbol can stand for something real, and because there are an infinite number of symbols possible for any real need, the symbolic needs are infinite in number. Therefore, any needs analysis is the search of a nearly infinite set of needs, and requires much imaginative thinking.

Often the potential users do not recognize their needs, partly because so many are symbolic. Advertising creates a new "need" for a man to express himself with a new car. New foods are ad-

USER NEEDS CATEGORIES ↓ DESIGN DIMENSIONS →	PERFORMANCE P, ΔP/ΔT	TIME T	COST C, ΔC/ΔT
PHYSIOLOGICAL NEEDS for air, water, food, shelter, sleep, exercise, reproduction and the security of their continuance (includes economic gains from efficiency and human body characteristics).			
SOCIAL NEEDS for conversation, friends, belonging to a group, group identity, social acceptance, etc.			
PSYCHOLOGICAL NEEDS for a sense of worth, accomplishment, autonomy, recognition, etc.			
SELF-FULFILLMENT NEEDS for the realization of one's potential.			

Figure 2.2. The Matrix Analysis of User Needs.

vertised until people feel that they cannot satisfy their hunger with calories alone. They "need" a cruncy mouthful. You may even feel the "need" for a digital watch. For these needs to be created, some latent need must have existed. It is only developed and recognized when the new design is put into effect. Anyone who can predict which latent needs will become real needs has got it made. We simple designers just do our best to imagine all kinds of possibilities and then we put them to the test.

THE MATRIX ANALYSIS OF NEEDS

Figure 2.2 is a diagram of a matrix which you can use in an imaginative search for the user needs. Into each box you put what you already know about the user needs, and what you imagine they could be too, real and symbolic. What you do in the beginning is to try and put as many things into these boxes as you can—pages and pages if you can do it, because you are looking for elusive diamonds in a pile of rock and debris.

WHAT RESOURCES WILL BE EXCHANGED

The next step is to decide which of the needs can feasibly be met with the restraints on time and resources which will naturally appear somewhere. Here are some of the ways that you can make a rough assessment of the resources that will be exchanged for meeting various needs:

1. The economic gain by the use of the design. A cost-benefit analysis.
2. The cost of alternative ways of meeting the need. For example, material can be transported by a conveyor, or by trucks, or by rail. Likewise, the need for a status symbol can be met by a car, house, clothing, and so forth.
3. For very new design objects with high utility in the symbolic dimension, there may not be alternative ways of satisfying that

need. As you will see in the toothbrush example, some assessment of the monetary value of an electric toothbrush was arrived at by comparing an electric razor with a manual razor. This technique involves the comparison of one design object with another design object going through a similar transformation.

4. Another way to arrive at the resources that will be exchanged, is to determine the cost of a design solution plus profit and distributing costs. Then make a survey of the users to find out how many would buy it.

5. Another technique is to use the intuitive judgment of experts. If you can find an expert and believe that person, then take that opinion. If you can get a panel of experts and get them to agree, so much the better. A new product advisory board is an example. There is a way of quantifying intuition and bringing about a consensus of a group of experts, and this is an excellent technique for getting a handle on what resources would be exchanged for a completely new product design. (See Appendix II.)

HOW MUCH NEEDS ANALYSIS SHOULD BE DONE?

When is an analysis of users' needs complete? Never! You can never know all there is to know about user needs. Nevertheless it is necessary to terminate the needs analysis and get on with the business of designing, making, and putting the object into place. Consequently, when designing, the majority of users' needs are assessed and the exchange of resources is estimated. At this point one proceeds with the design until more is known about it. There comes a point in the design process when it is generally possible to test users' needs. With manufactured products, it is desirable to have a pilot prototype and a pilot run phase. These "tests" aid in the reassessment of the estimated users' needs before any commitment is made to full production. In the case of one-off designs, such as a conveyor, it is not possible to do this. However, scaled-down models can be used with reactive panels to test satisfaction of users' needs before the final commitment of resources to heavy construction. The act of testing or reconfirming the users' needs is part of the principle of iteration. It is important to uncover helpful information early in the design process so that major iterations are avoided later in the process when they will be more expensive to put into effect.

APPLICATION TO AN EQUIPMENT DESIGN

This equipment example makes a technical transformation of energy, and as such this example will serve as a model for other equipment like a gear box, a materials conveyor, a pump, an elevator, and so forth.

Analysis of User Needs for a Substation Transformer in the Requirements Definition Phase

DESIGN DIMENSION	Technical Requirements
P	*1,000 KVA, 13,500/220 volts, 3 phase, 50 Hz*
$\Delta P/\Delta T$	*tropicalized*
	Other Needs:
P	*safety, truck transportable*
$\Delta P/\Delta T$	*convertability from 50 to 60 Hz*
P	*overload capacity*
C	*price $X*
T	*on-site in Y months for completion of transmission lines*
P	*serviceable by local tradesmen*
$\Delta C/\Delta T$	*maintenance and replacement cost under 5% per year*
	Resources Exchanged:
$\Delta C/\Delta P$	*This would be approximately $X per KVA, plus extras, determined by world prices on transformers. No reasonable alternative exists. Value is related to cost saving on protection from a lower voltage.*

In the Detailed Design Phase
the Technical Requirements
will be Expressed More Precisely

The need to have "overload capacity" will be pinned down by a customer-required test specification which may include a statement like the following:

"At an ambient of 30 ± 2°C, and an oil temperature of 55 ± 2°C, increase the load current to 150% of full rating at rated voltage. After 15 minutes the fuse should not have blown. At 60 minutes the fuse (or breaker) should have opened the circuit. Oil temperature should not exceed 100°C. Insulation resistance should be within 1% of its value at the start of the test, etc."

The need for completion in Y months will have been modified by circumstances that change the requirement—Needs can change. Suppose for example, it was found that shipping by sea would take one month more than planned. During the detailed design phase a schedule may be developed which looks like this:

Decision on oil and insulation materials	January 1
Design drawing sign-off and begin fabrication	March 1
Factory test of completed transformer	May 1
Customer commissioning test	May 15
Packed and shipped (one month earlier)	June 1
Arrival at destination in Latin America (Y)	September 1

APPLICATION TO A CONSUMER PRODUCT

This kind of engineering design requires very little in technical sophistication. Success depends mainly on other factors. Convenience and appearance are important, a purely functional design is not enough. People have nontechnical needs which are just as important as functional utility, especially in the case of mass-produced products where the user can choose among several products.

This example will be useful as a model for products like cars, TV sets, appliances, gadgets, portable calculators, hifi components and other products where success depends on the astute selection of technical and nontechnical needs to be met, and the user is free to use or not to use.

Analysis of User Needs for an Electric Toothbrush

DESIGN DIMENSION	USER NEEDS AT THE CONCEPT PHASE	
P	*Physiological Needs:*	*Clean teeth better than a hand brush, massage gums, reduce decay, hygienic family sharing, electrical and mechanical safety, etc.*
P	*Social Needs:*	*Sweet breath and white teeth (symbolic needs for social acceptance); handle colors to match bathroom, etc.*
P	*Psychological Needs:*	*Autonomy in deciding when and how one's teeth are to be cared for, self-esteem from care of teeth, praise for effort, pleasure from giving or receiving a gift, etc.*
P,ΔP/ΔT	*Technical Needs:*	*Diameter, length, brush size, amplitude frequency, weight, running time, reliability, useful life, etc.*
T	*Time Needs:*	*Needed for Christmas market.*
C	*Resources Exchanged:*	*$1 per person is the lowest cost alternative, but electric razors sell for twenty times the price of a manual razor, so probably $20 will be paid for an electric toothbrush.*

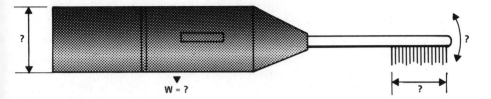

Figure 2.3. Consumer Product (Electric Toothbrush) Which Has to Undergo User Needs Analysis.

Some Differences in the Detailed Design Phase

At the beginning of this phase there should be a set of target speci-fications which reflect the previous market surveys and prototype testing. For example, the psychological need for "autonomy in deciding when and how one's teeth are to be cared for," and the physiological need for "hygienic family sharing", will boil down to statements like the following:

- There shall be six interchangeable brushes in distinctively dif-ferent colors.
- The case shall be adaptable to decoration and shaping to simu-late popular cartoon animals so that additional sales for children can be encouraged.

These statements are far from the detailed production specifications which will eventually result, and thus are a challenge to the mechani-cal engineers who must use their ingenuity to design multiple insert injection molded dies for the plastic cases—not a trivial design task.

APPLICATION TO A LARGE-SCALE
CAPITAL PROJECT

This example of a hydroelectric dam is the kind of design where there is much interaction with society as a whole. Many people are affected. Individuals do not choose whether or not they will use it. Success depends on a collective acceptance or rejection by large groups of people. As such, this example can be used as a model for other capital works like highway systems, a manufacturing plant, an oil pipeline, a computerized management information system, a transportation system, a shopping plaza, a community plan and so forth—wherever the social impact is significant to success.

Analysis of User Needs for a Hydroelectric Dam for Latin America at the Feasibility Phase—an Overview

DESIGN DIMENSION	GENERAL NEEDS:	
P	Technical needs for electric power, irrigation and water transport.	
P	Physiological needs for security of food supply by irrigation.	
P	Psychological need of accomplishment for the economic planners.	
P	Psychological need of safety through street lighting.	
$P, \Delta P/\Delta T$	Social needs of new communities created after flooding.	
P	Cultural needs of nationalism.	
$\Delta P/\Delta T$	Dynamic needs for growth and maintenance of design.	
P	Symbolic need for status symbols.	
C	Amortized cost not to exceed cost recovery by sales and taxes.	
T	Approval of world bank before next election of sponsor government.	
	ANALYSIS OF NEEDS:	
	System Outputs	Corresponding Exchange of Resources
P	500 megawatts electric power	land, capital and 2¢ per kwh.
P	100,000 inch-acres of irrigation water	$1 per inch-acre above dam

P	water transport above and below dam	water valued at $1 per inch-acre
$\Delta P/\Delta T$	reserve of water for dry years	$0.1 per inch-acre per year
P	standby electric street lighting	a local tax increase of 10%
P	community sites	$100 per acre
P	use of local materials and labor	training and inspection cost, 10% over local cost
P	attractive landscaping and lighting of dam	about 1% of national park budget
$\Delta P/\Delta T$	expansion of 100 mega-watts per decade	50% of saving in final installation cost
P	symbolism for national pride	public subscription for costs, about what was raised for a national opera house

Resources exchanged would be the present value of revenue from all outputs, taken over a fifty-year period. Note that some of the intangibles are estimated by a comparative method. These will be dominated by the highly valued technical outputs of power and irrigation water—unless political factors dominate.

Some Differences in the Detailed Design Phase

The requirements will be more precisely known and modified by practical considerations of cost. The original list of needs can never all be met. For example, after analyzing several possible dam sites, it may have been found economically practical to go to road transport in place of water transport. Thus, in the Detailed Design Phase, the engineers may be trying to meet needs defined more precisely by statements such as:

- The initial electrical power output required is 475 ± 25 megawatts.

- The irrigation water above the dam at the chosen site should not be less than 10,000 inch-acres in any month, based on rainfall tables of the past ten years.

Of course, the requirements would fill volumes, and they would probably be included in a very detailed design contract given to a firm consulting engineers. It is not likely that the aggregated needs of the users would change suddenly, although political changes could cause renegotiation of the contract to favor one need over another.

APPLICATION TO BUILDING CONSTRUCTION

User Needs (Requirements) for Cafeteria Building at a Petrochemical Plant

- *Facility to provide hot meals to 250 persons per shift with 45 minute lunch periods staggered over two hours.*
- *TV room for 50 persons with lunchboxes or light snacks.*
- *Games room for 50 persons with lunchboxes or light snacks.*
- *Kitchen to prepare hot meals on site.*
- *Separate visitor dining room with separate washrooms.*
- *Needed in two years when the operating staff start training.*
- *Value to company is about 10% of annual payroll of 5 million dollars.*

Comparison of Consumer Products with Industrial Equipment

In general, consumer products, as well as architectural works, must meet many human needs, with special attention to symbolic needs. On the other hand, industrial equipment, industrial processes, energy sources, etc., emphasize the technical requirements, with human needs being indirectly included in the performance specifications. All generalizations have exceptions, and that is the reason why a thorough search of needs is advisable. The analysis of needs makes possible an accurate definition of the design problem, its system, and the criteria by which to judge solutions.

CASE HISTORIES

A Material Conveyor—Esthetics
and Reliability

A public utility told their consulting engineering firm that they would like to have coal moved from the dockside to a stockpile which was about a mile from the lakefront. Before calling in vendors of conveyors, the consulting engineers attempted to define the requirements in some detail. Conveyors were not new and no development was needed. What was required was a very specific technical specification which could be put out for tender.

From the plan of the overall site it was noticed that the conveyor must pass a public roadway. It could either pass over or go under this roadway. The lowest cost was to go over the roadway, but as it was not far from a park area and also on a road which tourists used, there was some concern about the appearance of the conveyor. It is usual to transport coal on an open conveyor, or on one which has, for the most part, a shelter above the belt to keep the snow from fouling up the works. Such a conveyor is rather unattractive from the public viewpoint, although it functions perfectly well.

In this situation any proposal for a design had to be passed by the architects, so the consulting engineers talked to the architects first. From them they found out that the management of the public utility was very concerned about the appearance of its new generating plant, because of the difficulty in obtaining additional sites elsewhere in the country. In other words, it was realized that even though people needed the power, they did not like any unsightly public work to go along with it, nor did they want any dust. It appeared that the conveyor had to be well covered so that the wind would not blow coal dust into the neighboring community or on to cars going by. It would cost a little bit more to completely enclose the conveyor, but the management of the public utilities was prepared to pay extra for the sake of esthetics. As a consequence, the architects and consulting engineers agreed on the design of a fairly smooth-looking covered conveyor which ran from the lakefront over the roadway and back out of sight into the storage area. The structure was designed to harmonize with the buildings and not to be unduly distracting to people using the roadway. The waterfront portion was already going to be well screened by trees, in any case, so that open-truss conveyors could be used in this portion. The design solution appeared like that shown in Figure 2.4.

What about the technical requirements? The consulting engineers determined how many tons of coal per hour had to be moved. It was a tradeoff between the economics of emptying the ship very rapidly and the cost of a much larger conveyor than the consumption rate required. They also looked into the maintenance and reliability requirements. Because of the large stockpiles and the community noise requirement to shut down every evening, it was possible to do maintenance without seriously interrupting the unloading of the ship. However, a major breakdown would entail extravagant shipping charges, and therefore a means was provided for workmen to be able to walk along the conveyor at both sides

Figure 2.4. They Needed a Conveyor with Acceptable Appearance.

and inspect the rollers for over-heating or other indications of bearing failure.

With regard to the more or less continuous operation, one new requirement was made: There should be no supporting structure within 50 feet of the roadway in case it would be hit by a truck and the conveyor put out of commission. This made one of the spans unusually long. In all cases the client was prepared to pay the premium price.

A Beverage Can Opener—Human Factor Measurements

Before the days of tab-opening beverage cans, it was common practice to punch one or two holes in the top of the can and drink from it. An entrepreneurial group of designers felt that there was a need for something better on the market in order that the can might be opened with ease.

From a survey of competitive efforts, it appeared that the technical need was fairly well filled by existing beverage can openers. However, further investigation showed there was none that a child could use easily. Most adults would be happy to let a child open a beverage can if a manageable device existed. It appeared that a need existed for an economical opener which a child could use easily and safely.

The next investigation of need concerned the human factors. How much force could a child exert on a device?

From E. J. McCormick's book, *Human Factors Engineering,* (McGraw Hill, 1964), they took the following information:

Girls squeeze with one hand:

8 yrs.—10 lbs.
9 yrs.—15 lbs.

This information was inadequate for squeeze lever can openers, so one of the designers had his own children participate in an experiment in order to collect data. His daughters could squeeze with both hands:

4 yrs.—12 lbs.
8 yrs.—15 lbs.
9 yrs.—30 lbs.

In this example, the human-factor needs were transformed into technical needs. What this project needed, in addition to a performance analysis, was a study of the performance and cost needs as they would change with time. As you well know, tab-opening cans are a common commodity now and the beverage can opener would have had a very short market life indeed. Not that this was all bad, because if they had put it on the market quickly and it had met an existing need, it could still have been a successful design.

CONSOLIDATE YOUR KNOW-HOW

In this chapter you learned how to use the Matrix Analysis of User Needs for developing a long list of needs that may be met with a design concept. You then learned that it makes sense to meet only those needs for which resources will, or would be exchanged. That then, was the Principle of User Needs.

If you review the applications and case histories in this chapter, you will know how to go about applying your knowledge—but can you really do it on something of your own? Your answer will be found by doing some of the exercises that follow. Follow through and become a pro! Faciendo dicimus—we learn by doing.

DESIGN GUIDELINES

GUIDELINES FOR ANALYZING NEEDS.

Do	**Don't**
Do make sure that the engineered design meets the basic needs of its users.	*Don't ride your hobby horse of technical sophistication just to enhance your reputation as a top-notch designer. Let your ego get its kicks from the satisfactions expressed by the users.*

Do take a look at the resources that the user will or would exchange for what you will deliver.	*Don't forget the social and psychological satisfactions that the users may appreciate by paying well for them.*
Do take a look at the timing from the user's point of view.	*Don't start something you can't finish in time to be worth the effort. Better never than too late.*

Questions to Help You Consolidate Your Understanding of this Chapter

1. It is sometimes said that a good artist does his own thing and doesn't care what people think of his work. Now, consider the engineering designer: He could create just to please himself or he could create to please others. What, then, is the rationale for meeting user needs?

2. Ask people what they need and they want everything. Yet there are always limits on time and money for a design effort. How does the designer make decisions on what is to be served and what is not to be served by a design?

3. The needs that are met by a passenger automobile are many. Where do the following car items fit in the user needs categories of physiological, social, psychological and self-fulfillment? (The needs may fit into more than one category.)

- Steering wheel
- Heater
- Chrome on bumpers
- Front grill
- Back seat
- Tape stereo deck·

Questions on the Case Studies

1. For the case history of the material conveyor, suppose that the public utility had not thought of the esthetics of the conveyor section crossing the road. After all, the cost of energy production

must be kept down, an open truss conveyor is the cheapest, and in addition the user is the utility and not the public. What are some of the possible consequences of that course of action?

2. For the case history of a beverage can opener, the public were free to use it or not to use it. If the idea of meeting the need for children to use the can opener doubled the cost and put the selling price over $5, what information would you want in order to decide which way to go?

Assignments to Help You Become a Pro

1. If you are practicing your skill on a minor design project as suggested in chapter one, do a needs analysis for it, indicating where you would require extra information and how you would go about getting it.

2. Take an engineered design that you are familiar with. Find the nearest model in this chapter under the applications to a product, equipment or a large scale capital works. (If you do not have a design that you are familiar with, choose one mentioned as similar to the applications.) Do a matrix analysis of user needs, then estimate the resources that will, or would be exchanged for meeting them. Sum them up and compare with the selling price or cost of an actual engineered design.

3 Defining the
Project Objectives

KEY IDEA: THE DESIGN EFFORT SHOULD BE GUIDED BY WORK-
ABLE OBJECTIVES RIGHT FROM THE BEGINNING.

GETTING A FAST DESIGN SOLUTION

Here is a design problem that I have given to many groups at my
seminars on engineering design methods.

"A movie actress has a problem with her house in Beverly Hills.
She complains that the bathroom mirror gets all steamed up when
she has a shower. Naturally, she needs the mirror right after the
shower. You are an engineering friend and have recognized a
technical need. You have offered to design something that will
correct the problem.

Some of the technical aspects of the problem are as follows:
our actress friend takes long, hot showers which raise the bathroom temperature to 95°F, while the mirror remains at a wall
temperature of 75°F. The mass of the 1 meter X 2 meter mirror
is 25 Kg. The thermal capacity of the glass is 100 Watt-sec. per Kg.
The cooling coefficient is 5 Watts per sq. meter.

What do you suggest be done about the problem?"

When a group of design engineers is given this problem, the following happens: about 75 percent go immediately into the calculations

45

needed for the design solution of a heated mirror. After progressing in this direction a way, some of the other engineers will ask the question, "What are we *really* trying to do here?" They have a hard time making themselves heard because three-quarters of the people are busy on the engineering solution, but eventually they or I get through with this question. Then the design effort moves to other solutions which are quite feasible, although less sophisticated technically.

The new solutions then include such things as moving the mirror, installing another mirror in an adjacent anteroom, exhausting the vapor from the room, putting a fan on the mirror, putting an enclosure around the shower area, and so forth. The search for creative solutions begins only after they recognize and describe the basic design objectives.

**WHAT YOU WILL GAIN FROM
READING THIS CHAPTER**

Objectives

1. *You should learn how to set design objectives that are comprehensible and workable.*
2. *You should learn how to establish tradeoff guidelines.*

Benefits

1. *Good design objectives will encourage you to search for creative design solutions.*
2. *You will be able to control the drift away from the targets for performance, cost and time.*

When do you set objectives? You should set or recheck the design objectives of every new design immediately after checking the analysis of needs.

In this chapter, you will first reason your way through to the Principles of Setting Objectives, then you will learn what constitutes a comprehensive set of objectives for performance, cost and time, and for their tradeoffs. To make these ideas real for you, various applications are given; one for a design which is mainly technical; one for a consumer product, and one for a large-scale capital project. These are followed by the case histories of objective settings for a recreational camper, a house, a garbage disposal system, a prosthetic device and an ice-breaker. These are critically reviewed so that you will learn how to recognize a truly comprehensive and workable set of objectives.

THE PRINCIPLE OF SETTING OBJECTIVES

An extensively used management technique is MBO—Management By Objectives. By this means, the efforts of many persons can be guided to obtain the desired results. This powerful technique of management can be used to advantage in motivating and controlling the design effort to get the desired results. However, because engineering is more complicated than making and selling breakfast foods, we will need some elaboration of the idea of setting objectives.

An engineered design is a physical reality which brings about a change in a state of affairs. The new state can be defined in advance by objectives and target specifications.

Let the present state be defined by a vector, S_1, which is composed of the affected variables, V_1, V_2, V_3, . . . to V_n which are set at their present state. When the design is completed and in place, it can be defined by a new state vector, S_2, with the variable set at new levels. However, because the design usually involves the placement of a new object, there will be new variables included in S_2. Therefore, we can write that S_2 = the states of the NDV (new design variables) + (the changed state of the variables in S_1). Thus,

$$S_2 = NDV + S_1{}'$$

where

$$S_1{}' \equiv \overline{V_1, V_2, V_3 \ldots V_2} \text{ in the new state}$$

Another way of looking at the new state is to consider that S_2 is equal to the *how* plus the *what*. The *what* is the $S_1{}'$ which is what is achieved by the design solution. The *how* is the NDV or the actual state of the variables of the design solution itself. To those familiar with the systems concept, Figure 3.1 represents the present state and the new state.

Consider an example such as a design of a sun shelter. The initial state, S_1, includes the dimensions of the area to be sheltered, the condition of the soil, and any objects that may be in the way. It also includes the environmental factors such as temperature, the attenuation of the sun's rays, attenuation of wind, etc. Now if we design a simple shelter which blocks the sun's rays and the outside air, we have changed the environmental variables on the site and also some of the dimensions. We have changed S_1 into S_1'. We have also included new materials which have dimensions. The states of the new design variables, NDV, define the design solution, and these are included in the new state, S_2. Therefore

$$S_2 = NDV + S_1' = \text{sun shelter} + \text{changed environment}$$

Now, if we go back to looking at the variables in S_1', which state *what* is to be achieved, then there may be many ways of achieving this. If NDV, the new design variables are left open-ended, we can

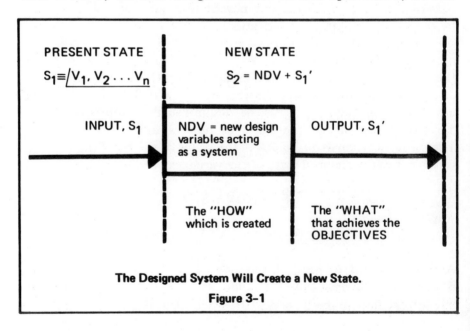

PRESENT STATE

$S_1 \equiv \underline{/V_1, V_2 \cdots V_n}$

NEW STATE

$S_2 = NDV + S_1'$

INPUT, S_1

NDV = new design variables acting as a system

OUTPUT, S_1'

The "HOW" which is created

The "WHAT" that achieves the OBJECTIVES

The Designed System Will Create a New State.

Figure 3-1

look for creative ways to achieve the change in S_1. In our example there are many kinds of shelters which will give protection from the sun and the weather elements.

In many technical designs, if we achieve the *what* by designing something that provides S_1', we have a design solution. For example, a conveyor will take a pile of coal, S_1, and transform it to a new pile, S_1'. The case of an automobile is not so simple. A set of specifications will define S_1', but will it be an acceptable solution? Maybe. It is better to say something about the purposes served by S_1'. These are called objectives. They state what is to be achieved when the new design is in place. (For more on the relationship of S_2 to objectives, see chapter eight on decisions and chapter ten on optimization.)

For an automobile there are technical objectives pertaining to the accommodation, speed, economy—but equally important are objectives pertaining to esthetics, comfort, convenience and symbols. It is convenient and advantageous to first establish descriptive objectives which state what is to be achieved, and then follow them with target specification which are measurable details of S_1'. This two-step description of the new state is extremely valuable in providing guidance for the tradeoffs which will be necessary. The higher level descriptive objectives should not contain the solution. We concentrate on the *what* and not the *how*. We are now ready for a statement of the principle.

THE PRINCIPLE OF SETTING OBJECTIVES

When the design objectives state **what** *is to be achieved rather than* **how** *it is to be done, the possibilities for creative design solutions are enhanced.*

OBTAINING GOOD OBJECTIVES

Good objectives arise out of our needs analysis (chapter two) because in that step we looked at what the users will exchange in resources and we decided to meet certain needs. In other words, *an objective can be thought of as a statement of "meeting the following users' needs."* So, in the shelter example, an objective is "to meet the user needs of protection from the sun, the rain, and the wind."

This is a statement of *what* is to be achieved and not *how* it is to be achieved. It also is oriented towards what the users need. (see chapter two)

In the early phases of design, that is, in what we call the upstream design activity, we are setting target objectives because we do not know exactly which objectives can be met or to what degree they can be met. As we move downstream in the design process we get a better idea of what specific design will meet the needs of the users. Therefore, it is commonplace for downstream objectives to be very specific and to contain statements about the design solution itself. In other words, the objectives change from an orientation of *how* to an orientation of *what*.

For example, in the construction phase of the sun shelter, there are specifications that must be met and these become incorporated into the objectives. This is the way design descriptions progress, but there is one danger: *the design solution may be too early a substitute for the design objectives.* Downstream objectives are only complete if they continue to contain the statements of what is being achieved by the specified design solution. This is necessary because there will still be minor tradeoffs that must be made during the construction or implementation phases and these should be guided by the objectives.

THE THREE DIMENSIONS OF OBJECTIVES

Design objectives can be cast in a framework of three dimensions which are independent in themselves, but which become related in a design object.

Performance Objectives, P

These are the attributes of the design solution when it is in place and working. These include all the functional performance parameters of the design solution plus those appearance and convenience aspects which are important. The performance objectives will state what is to be achieved by the presumed unknown, $NDV + S_1'$.

Performance objectives include all of those things except time and cost.

Time Objectives, T

These are the target values for a design schedule. They reflect the best estimates of the time required either to obtain a solution in the shortest possible time or, better still, the time when the users would best be served by the design solution.

Cost Objective, C

These objectives include all of the investment of material and human resources in preparing the design and for the materials and labor of constructing and putting it in place. All monetary considerations can be included in the cost objectives, even those pertaining to revenue and profit, when applicable.

A comprehensive set of objectives should include statements about what is to be achieved as target performance, as well as what target amounts of resources or cost are to be assigned to the project. Any set of objectives which does not include all of these three major dimensions is lacking in one important aspect and is likely to guide the design effort to an unbalanced result.

At the beginning of any design project you will find that either performance, time, or cost will dominate. If the problem is technically difficult, performance will dominate over cost and time. On the other hand, the solution parameters may be fairly well known and the time or the cost will dominate.

Given that Performance, Cost, and Time (P, C, T), are in the satisfactory region, then at any point of the design process, one of these will dominate over the other two for major and minor tradeoffs. This will be self-evident to you if you consider design projects one at a time. However, as a design moves downstream in the design process, the relative importance of the P, C, or T objectives can change.

In the beginning, we generally find that P dominates, and in the end we find that T or C will dominate. The reason for understanding this will become clear in chapter five when we talk about tradeoff controls. In the meantime, suffice it to say that the minor tradeoff increments, ΔP, ΔC and ΔT, can usually be rank ordered, with the provision that P, T and C be within the satisfactory region. This will guide the minor tradeoffs that invariably will be made throughout the design process. You can see how this is done in the applications of this chapter and in chapter five.

IDENTIFY THE CONSTRAINTS

The setting of target objectives which state what is to be achieved is a large part of a good problem definition. These objectives can generally be satisfied in many ways. What will eliminate otherwise satisfactory solutions are the limitations caused by the constraints. Almost every design problem must be solved within certain constraints, and therein lies the challenge.

For example, a design solution in one part of the country may be unsatisfactory in another part of the country because of different environmental or legal requirements. These are constraints. Mathematically speaking, the variables in S_2 must have states which do not exceed certain predetermined limits. For an example, a stove or refrigerator must fit through a door opening. These are dimensional constraints on the design solution. The new state of the variables, V_n in S_2, may be constrained by limits such as $V_{max} > V_n > V_{min}$.

The way to identify constraints (restraints) is to look in the following categories:

Dimension Constraints

Many designs have limitations on their physical size. For example, a railroad car must fit existing tracks; major household appliances must pass through a doorway; a retrofit replacement part for an aircraft must fit in exactly the same place that the previous part occupied. It is not only the length, breadth, and height dimensions, but the openings for fastenings, the thread size for interfacing with adjacent mechanical parts and so forth. When a design cannot be as big as you want or as small as you want, then there will be dimensional constraints.

Laws of Nature

Engineers are supposed to know about the laws of nature. Plastic will melt at a certain temperature and is, therefore, banned from use in certain applications. The laws of physics state that pressure times volume over temperature equals a constant, and when you design tires for an automobile you violate this law only at your risk. Each engineer tends to know those physical laws within his discipline but does not know much about those in other disciplines. The mechanical engineer will often get into trouble by violating a law of chemistry and vice versa. This is why multidisciplinary cooperation is necessary.

Laws of Man

There are legal constraints on most designs. For some, there are the Underwriters' Laboratory constraints, for others, there is the Canadian Standards Association. Sometimes there are trade association standards which have been agreed upon. There are many for the safety of individuals and society, and quite a few new ones pertaining to environmental pollution. Engineers who work regularly with one type of product or design activity generally know about the legal restraints, but should be cautious if they move into another kind of design activity.

Social and Cultural Constraints

Many of our restrictive environmental and safety laws pertaining to automobiles and other equipment are the result of social and cultural pressures which eventually resulted in restrictions by law. These pressures precede the legal restrictions, so effective designers anticipate these requirements. For example, we fully expect there to be legal constraints on the efficiency of future household appliances and passenger automobiles. Smart designers are setting themselves limits on energy efficiency in anticipation of new regulations.

Man-imposed Constraints

Sometimes the client, customer, or sponsoring organization will dictate limits or constraints on the design. They don't want this or they don't want that—at any price—and the designer is compelled to accept these as constraints. Acceptance of them, and including them in the problem definition, makes them visible, and if they are unreasonable then they may be removed or modified by the person who imposes them.

Special Difficulties

Something like constraints are those special technical difficulties that must be solved in order for a design to be successful. For example, in order to complete the design a group of engineers may depend upon the successful outcome of a piece of research which defines a parameter for them. If this is the case, and it is outside their control, it is one of the special difficulties that must be overcome. It is best to enumerate this kind of special difficulty as if it is a constraint. The idea is to point out special difficulties that must be overcome in order for the design to succeed.

APPLICATION TO AN EQUIPMENT DESIGN

This equipment example represents a technical transformation of energy. It will serve as a model for other equipment like a gear box, a materials conveyor, a pump, an elevator and so forth.

Because this is mainly a technical design, the objectives will very closely resemble the quantified needs listed for it in chapter two. However, as is often the case, not all needs can be met because of the usual limitations on time and money.

The Objectives for an Electrical Substation Transformer in the Phase of Definition of Requirements

DESIGN DIMENSIONS	OBJECTIVES
p	*To handle 1000KVA and transform 13,500 volts to 220 volts at 3 phase 50 Hz*
$\Delta P/\Delta T$	*To be convertible to 60 Hz with less than 5% change in rating*
$\Delta P/\Delta T$	*To be tropicalized for use in the Caribbean*
P	*To have an overload capacity of 50% for 30 minutes*
P	*To be transportable by truck from ship to site*
P	*To be serviceable by the local utility electricians*

C	*Installed price to be $X or less*
$\Delta C / \Delta T$	*Maintenance cost to 5% per year or less*
T	*To be completed and installed in Y months so as not to delay other parts of the power system*
	Tradeoff Guidelines
$C > T, P$	*As this transformer is from a standardized design and under contract for time and cost, no major tradeoff is expected. However, if unusual circumstances call for a tradeoff, we would favor proposals extending the completion time rather than increasing the cost or lowering the performance because the customer has limited funds and the performance standards are partly determined by government regulations for safety.*
$\Delta C > \Delta T, \Delta P$	*For the marginal tradeoff, given that performance, cost and time are in the satisfactory region, we would prefer marginal tradeoffs that reduce the cost, for the same reasons as above.*
	Constraints
	Some constraints are the industry standards for safety, government safety regulations, and the road clearance available for transporting the transformer.

Comment: The technical objectives will more than likely remain the same throughout the design phase. However, the time objective may change if slippages occur in the installation of other parts of the power system, for which this transformer is only one component. At the production phase, the objectives will be augmented with a full set of drawings and specifications.

Application to a Consumer Product

This example will be a useful model for other applications of consumer products like cars, TV sets, appliances, portable calculators, stereo components, and other products where success depends upon meeting both technical and nontechnical objectives.

If you will refer to the chapter two example of the user needs for an electric toothbrush, you will see that not all the "needs" are incorporated into the objectives. Consumers want everything. When you set comprehensive objectives by analyzing what resources will be exchanged in order to meet needs, you exercise a degree of judgment. This requires a skill that few have. You should test your judgment by consulting experts and consumers.

The Objectives for an Electric Toothbrush in the Preliminary Design Phase

DESIGN DIMENSIONS	OBJECTIVES
P1	*To be attractive as a toothbrush, suitable for sale primarily in the gift market, and secondly as a personal purchase.*
P2	*The technical functions are to be at least as good as past "family" models of brand X.*
C	*The selling price is to be within 10% of present utility models.*
T	*The time from design-start to production-release is to be about one year.*
	Tradeoff Guidelines
P1 > P2 > C > T	*As the market for this product is already served by competing models, the attractiveness is of paramount importance. Should a major tradeoff be necessary we would favor the attractiveness over the technical functions, over cost reduction, over the timing.*
ΔP > ΔC > ΔT	*At this early preliminary design phase in the design of this electric toothbrush, we are not yet pressed on time or cost, so the marginal tradeoff should be in the same direction as the major tradeoff.*

	Constraints
P	*Must meet UL, CSA, and dental safety standards.*
C	*Tooling cost not to exceed the $50,000 available.*

Comment: First, some of the technical needs outlined in the previous chapter have not been enumerated. We will use these in the next chapter for criteria which measure the accomplishment of P2.

Second, the major tradeoff and the minor tradeoff guidelines are the same. This is because the example is for an upstream case. You could well imagine a change just before the production phase where the major tradeoff guideline would remain the same, but the minor would change. Time is pressing, good enough is enough. Whence, at that point of time, the design team could be guided by: $\Delta T > \Delta C, \Delta P$.

APPLICATION TO A LARGE-SCALE
CAPITAL PROJECT

Large capital works like a hydroelectric dam, a highway system, a manufacturing plant, an oil pipeline, a community plan, and the like, have some things in common.

1. They are one-of-a-kind.
2. Experimental prototypes are not possible.
3. They tend to involve the application of a well understood technology.
4. Ultimate success for the project depends on its acceptance and careful use by many people.

Hence, objectives will be both technical and social, although the social ones are often included in the technical requirements, e.g., the need for the safe use of streets to visit neighbors during the evenings would be included in the total electric power which includes the street lighting.

The Objectives for a Hydroelectric Dam for
Latin America at the Feasibility Phase—
An Overview

DESIGN DIMENSION	OBJECTIVES
P1	To produce 400 megawatts of electric power at start.
P2	To produce 500 megawatts one year later.
P3	To provide 50,000 inch-acres of irrigation water.
P4	To provide an irrigation reserve of 25%, with more being available by power cutbacks.
P5	To provide water transport below and above dam.
C1	To use local materials and labor.
P6	To provide an object of national pride.
C2	To be installed and operating for a cost of X millions of dollars.
T	To be completed by January __, 19 __.
$\Delta P / \Delta C$	Expansion of 100 megawatts per decade at cost Y per decade.
	Tradeoffs
$P, T > C$	At the feasibility phase, we do not know what the final costs will be until we have engineered the potential sites. As the client is a new oil exporter, we think they would favor major changes in the objectives, if necessary, which favor the performance and time over the cost.
$\Delta T > \Delta P, \Delta C$	The minor tradeoff guidelines are determined by our ability to get to the client fast with a contract proposal for the detailed design. Internally, therefore, we favor the incremental time over the quality level and cost of the study, provided of course, they are within the satisfactory level.

P,C C,P	Constraints *Foreign aid contract restrictions.* *Maintainable by local personnel.*
P	Special Difficulty *Must preserve national shrine in potential flood area.*

Comment: Naturally, there may well be some changes in the objectives and tradeoff guidelines as we progress into downstream phases in this design. Either good luck or bad luck may come from a site selection. The project will take many years to complete, and like the Aswan Dam, the client country may experience political changes that call for changed design objectives.

APPLICATION TO BUILDING CONSTRUCTION

Objectives for Cafeteria Building at a Petrochemical Plant
  (as they would appear before preliminary design)

P1 – A cafeteria suitable for timely serving of hot meals to 200 persons per staggered meal break.

P2 – Fast service for beverages and light snacks.

P3 – A cooking facility which will enable economical and nutritious hot meals to be prepared.

P4 – A room with TV viewing for 50 people while eating from lunchboxes or eating light snacks from the fast service.

P5 – A games room where people engaged in routine work can have some excitement after eating. (This is less than the ideal need.)

P6 – Washroom facilities for use without delay by the expected traffic.

P7 – Public telephone service.

P8 – Provide place for company announcements.

P9 – Provide clothing and footwear storage while on meal break.

P10 – Provide for efficient and economical maintenance.

P11 – To be efficient in the use of energy.

P12 – Provide for an expansion of 25% in numbers without major structural alterations.

Tradeoffs: *If major tradeoffs be necessary, change proposals should favour time over performance over cost because not being ready as planned would be costly (T > P > C).*

During the Detailed Design Phase, if P, T & C are in the satisfactory region, the minor tradeoffs *guideline is to favour marginal improvements in performance over the marginal changes in cost and time (Δ P > C, Δ T).*

CASE HISTORIES OF OBJECTIVES
SET IN EARLY DESIGN PHASES

A Recreational Camper

A mobile shelter in a low price range is required. It should have a sleeping capacity of four. One person should be able to set it up or dismantle it in approximately 15 minutes. The shelter should also be easy to store, safe to use, rugged and compact for towing behind a car.

A House Design

1. The total cost of the building is to be in the $75,000 to $85,000 range.
2. Construction must be started in the spring.
3. The lot is situated on a corner 80 by 150 feet, is relatively flat and without trees, therefore the design cannot rely on nature for beauty.
4. Furniture for the bedrooms, dining room, and living room is already owned, therefore they must fit accordingly.
5. Already have expensive carpets 10 feet by 15 feet which should be considered in design.
6. The lot is completely serviced by sewage, but will not be hooked up for nearly four years; therefore, a septic tank must be installed.
7. Septic tank must be 25 feet from house and 10 feet from all lot lines.
8. Master bedroom could be built on at a later stage.

9. Have to buy refrigerator, stove, washer, and dryer as major appliances.

A Garbage Disposal System

We are concerned with a small rural town. At present it has an average daily refuse output of about 70 tons.

Our primary objective is to dispose of this refuse by the most economical method available, and also, to avoid creating any nuisances or hazards to public health and safety.

There are a few factors to be considered in selecting the best method of disposal.

1. The town has no incinerator.
2. The town is now using the sanitary landfill method, with the site located four miles from the city limits. The present site has a projected life of 10 years, but there is an option to buy the adjoining land, which will extend this life to 25 years.
3. The town is now responsible for the collection and disposal of the rubbish by weekly curbside pick-up.
4. The present cost of operating the site is $1.88 per capita. The cost of collection is $2.17 per capita.
5. The annual rate of population increase is 1900. This rate is expected to increase with the arrival of new industry.
6. Twenty-two percent of the population now lives in apartment blocks of fifty or more people. This figure increases slightly each year.

A Prosthetic Device for a Man with One Arm and One Leg

The solution should make this man mobile. By mobile we mean he should be able to perform all the necessary daily functions by himself. This includes such things as getting out of bed, fitting the device, dressing himself, washing and shaving. He should also be capable of walking, eating, climbing stairs, opening doors, and performing his daily job . . . The device should not make him too self-conscious of his handicap . . . The apparatus must not place any undue strain on the patient.

A Case History of the Objectives for the Icebreaker Project in the Construction and Acquisition Phases

To deliver to the Coast Guard two operational "R" Class Icebreakers to the requirements defined by specification and contractual documentation at a cost which is as far below ceiling cost of $30 million as possible and, within Design Change Cost Allowance not exceeding $2.5 million.

An operational "R" Class Icebreaker is defined as being suitable for icebreaking service within the areas defined in the specification, having in mind the rigorous demands resulting from climactic, shock, and vibrational extremes.

COMMENTS ON THE CASE HISTORIES ABOVE

Question 1: Do any of the above problem definitions contain the solution?

Answer: Although some of the objectives contain elements of a solution, they are all sufficiently general to allow some freedom to the designer. It must be remembered that a problem definition results from a vague idea of how some design will meet a need. Therefore, the problem definition tends to include the original design idea. A good set of objectives will broaden the range of possible solutions rather than narrow it. For example, the objective of the garbage disposal system is to dispose of refuse and not "to build an incinerator." The latter would contain a solution. For the icebreaker, the objectives are in a downstream design phase and it is known that an "R" Class Icebreaker will do the job and hence, the solution replaces the objective. Note that the basic purpose is still included in these objectives.

Question 2: What constraints are placed on the design in the problem definitions of the camper and the house?

Answer: The problem definition of the recreational camper is relatively open and does not contain the constraints of the design problem. These were found later by the designing group. There are legal constraints on width, height, lighting, and hitches which restrict the designer. We might see the "capacity for a family of four" as a constraint, but at this stage, it appears more as a target than as a limitation.

When looking at the house design, there are specific constraints such as the lot size and the contours and the septic tank clearances. The other items, such as price, starting time, use of existing furniture, and appliances, look like some criteria by which to judge the proposed solutions.

Question 3: What needs are to be met by the objectives? Should the camper provide outdoor comfort? Should the house provide for family shelter, privacy, recreation, etc.? What specific family needs are to be met?

Answer: In this respect, both sets of objectives tend to be quite specific, and include some criteria for a successful design solution but do not have enough general objectives. The objectives for the Prosthetic Device are a good example of objectives in terms of needs.

The objectives for the icebreaker appear to contain a constraint of $2.5 million on the Design Change Cost Allowance. This is a man-imposed limit. This statement really expresses a tradeoff guideline which could be paraphrased as, "Given that the cost is in the satisfactory region, marginal improvements in performance are permitted, but major ones are not.

In all of the case histories above, we are lacking time objectives; in some we are lacking cost objectives. (Real world design case histories are seldom perfect.)

BUTTON UP WHAT YOU HAVE LEARNED

By now you should know the way to go about setting comprehensive and workable objectives. You should also be able to state guidelines for the tradeoffs that are likely to be made. In other words, you state where the targets are and give guidelines on what to do if the design project is getting a little off the track, which is almost inevitable. Good tradeoff guidelines will enable you to control that insidious drip of time and cost. In addition, you will be able to use guidelines to get the project back on track if the need arises.

To really make this knowledge your own you should apply these concepts to practice projects or to one of your own projects. You do this by continuing on at the end of this chapter. At this point we will summarize what has been learned in a set of specific guidelines.

DESIGN GUIDELINES

GUIDELINES FOR SETTING DESIGN OBJECTIVES

Do	**Don't**
Do set target design objectives before you start finding solutions.	*Don't rush into a design effort and burn yourself out by working toward some hazy objective. Find out where the target is before loosing your arrows.*
Do have objectives that state what is to be achieved rather than how.	*Don't restrict yourself and others by objectives that limit the search for creative solutions. Maybe there is a better way after all.*
Do have a comprehensive set of objectives which cover performance, cost, and time.	*Don't leave it up to the engineers to fight it out with the accountants and schedule-drivers.*
Do get agreement on the major and minor tradeoff guidelines.	*Don't leave tradeoffs to chance unless you like overruns in cost and time.*
Do revise the objectives whenever the user's needs have changed.	*Don't wait until you are finished to find out that the users have changed their minds.*

FOLLOW-UP EXERCISES

Questions to Help You Remember
What You Learned in This Chapter

1. Name at least three attributes of a good set of objectives.
2. Give at least one good reason for the design objectives requiring a change during the life of a design project.
3. What is one useful way to control the balance of performance, cost, and time?
4. In this chapter, I did not explain the why and wherefore of the tendency for engineering persons to jump into the design process without clarifying objectives. Do you have an explanation? (This is a good exercise for a group discussion.)

Questions on the Case Studies

1. In the case of the prosthetic device, suppose that objectives for time and cost are added. Suppose also that this design effort is about to begin and that there is no design solution that is known to the designers. What tradeoff guidelines would you give, and how would you state them?
2. In the case of the icebreaker, the designated area of service was mainly in the St. Lawrence River between Montreal and the ice-free water downstream from Quebec City. When the design was completed and construction started, the engineering team proposed a major design change which enabled the icebreaker to be of service in the Arctic waters north of Hudson's Bay. This would have increased the total cost of the project beyond the targets, including that allowed for design changes. Would you say the proposed changes were contrary to the objectives for the design project? As you understand the objectives, what would you predict the outcome to be?

Assignments to Help You Become a Pro

1. Using one of the models in the application, develop a comprehensive set of objectives for your practice project or a real design project.

2. Apply the Guidelines for Setting Design Objectives and make any changes to your objectives that you think are necessary.
3. Submit your objectives to a critique from an instructor or colleague. If you are working on a real project, try and get opinions of the users, their representatives, or a substitute for the user's viewpoint.
4. Make a prediction of the way you think tradeoff guidelines will change for the various phases you have planned for your design project.

4 Pinning Down the Design Requirements

WATERING DOWN THE SPECS

Anyone can make good specifications for measurable parameters such as size, rpm, voltage, power rating, etc., with their corresponding tolerances and quality control levels. When these kinds of specifications are applied to design alternatives, there is no argument as to whether they are met or not. The majority of engineering specifications are of this type. *However, the specs that give us trouble are those which are difficult to quantify—such as convenience, producability, maintainability, texture, color, esthetics, etc.*

Consider what happened to the specifications at a large manufacturer of technical products. The initial target specifications (requirements) were drafted up shortly after the engineering team started to work on the new product. These were not completed and issued until the product was about to go into production. In other words, no one had formally stated what the design would do when it was completed—until whatever they had was ready for production. This was the "safe" approach for the engineers because it guaranteed that what they designed would meet the specifications.

67

It did not, however, guarantee that what met the specifications would meet the customer's needs. It is possible they lost sight of that point. However, assuming that the specs came close to the customer's needs, let us look at some of the difficult categories of their specifications.

Under the category of "Installation, maintenance and repair," there was a statement that said:

"The design shall provide ease of maintenance and installation in the field at a cost equal to, or less than, present practices on a cost per unit connection."

Two years later this is what it said:

"The design shall provide for ease of installation in the field."

In two years this specification had become less definitive. One could not discriminate between the satisfactory design alternatives and the unsatisfactory ones.

Next, let us look at a more difficult category, "Mechanical and physical requirements." One of the target specifications said:

"Appearance: the product shall be esthetically pleasing, as might be determined by subjective evaluation of opinions of representative population samples."

Two years later the specification was issued in this form:

"Appearance: The product shall be esthetically pleasing."

Again, instead of making the specification more discriminating, they backed off a mile. Unfortunately for the company, with these sloppy specifications, they got only what they asked for. The design was "easily maintained"—if you had a kit of special tools and expensive factory-built test equipment. Actually, it was cheaper to throw it away. As for being esthetically pleasing"—it pleased the engineers. The only trouble was that the competitor had a better-looking one. Enough said! There *is* a better way.

WHAT YOU WILL GAIN FROM THIS CHAPTER

Design objectives are great for pointing the way, but how do we know when we get there? We need measures so that there can be no doubt about hitting the bull's eye on a target objective.

Objectives For This Chapter

1. *You should learn how to obtain specifications from objectives.*
2. *You should learn about scales of measurement.*
3. *You should learn how to put objectivity into measurement.*
4. *You should learn the test of a good criterion.*
5. *You should learn to know the difference between target specs and final specs.*

Benefits From This Chapter

1. *You will be able to develop meaningful specifications for all situations.*
2. *You will be able to reduce the disputes over what is good enough for the intangible factors.*

The step, "Develop Target Specs" is logically done after you have "Set Objectives," as in the previous chapter. In practice, you will tend to iterate back and forth between them until you are satisfied with their consistency. Your specs will become more and more detailed as you proceed downstream, whereas, the objectives will not increase in size or scope.

In this chapter, we will look at the major aspects of specifications and generalize up to a principle. Then you will learn how to add objectivity to those intangibles that you would like to specify and control. You do this by learning about different measuring scales and their relative objectivity. You also learn of a way to add objectivity to something as intangible as esthetics.

The applications are samples of deriving specs from objectives for the examples covered in previous chapters. There are three models which cover a wide range of situations. Then there is a case history which shows you how some designers went about developing target specs. My critique on their effort shows you how to improve your own efforts. The best of this chapter is then summed up in Guidelines for Developing Target Specifications.

At the end are exercises to help you practice and become a super-pro.

THE PRINCIPLE FOR DEVELOPING
TARGET SPECIFICATIONS

Criteria Measure Achievement of Objectives

How do we know when the objectives are met? What standards do we use to select a design solution?

For example, if we desire a steam engine to have an output of X horsepower for an input of Y tons of coal per hours, and a design solution meets this criterion, then we could say, "Mission accomplished." However, judging the design solution is not that simple. There are generally many criteria which must be satisfied and some cannot possibly be satisfied at the same time as others. Tradeoffs must be established. For example, the steam engine will have objectives pertaining to weight, noise, pollution, cost, time of completion, safety codes, and so forth. Thus, we will need many criteria by which to judge the design solution.

In the beginning of the design process, these criteria are *target specifications.* Towards the end, they solidify into *mandatory specifications* for constructing or manufacturing the design solution.

Are the Criteria the Same as the Needs?

Do the design criteria reflect precisely the needs of the users? This might be so with the perfect design but perfection is never achieved.

An analysis of needs of toothbrushes may show that people "need" an electric toothbrush which would sell for 79 cents. Clearly, the design criterion would be different from the apparent need. We may establish it at $10, with a vastly different anticipated volume of sales than we would get at 79 cents. The point to remember is that the ***design criteria are deliberately established by the designer*** taking into account many factors (as if he were God). They are frequently somewhat less than the identified needs (or requirements) themselves since the resources and time for a solution are not infinite. For example, a transportation system may need to transport 100 passengers per hour at peak periods. A designer may deliberately choose to design with the criterion of 50 passengers per hour and tolerate the lineups during peaks.

How Many Criteria

It must be admitted that some very fine products have been designed strictly from vague intuitive criteria. Works of art do not lend themselves to the establishment of criteria. However, anything which is to be mass-produced or constructed by persons other than the designer, will require specifications in order that the end product does what it is supposed to do. A simple product like a can opener may only require a few lines of specifications and a drawing. A civil engineering structure may require a hundred pages of specifications and drawings. A space orbiting satellite may have enough specifications to fill an entire room. The important thing about specifications is not how many there are, but how *effective* they are.

As described in chapter three, before deciding on a design solution, *we set objectives which describe what is to be achieved.* These are descriptions of a new state of affairs. There are objectives for performances, cost and time. *How do we know when the objectives are achieved?* By criteria, that's how. Each objective will result in from one to hundreds of criteria. The more criteria we have, the better our expectation of having met the objective. Take into account that the new state of affairs is very complicated, having thousands upon thousands of variables existing in new states. No amount of objectives or criteria can completely describe the new state—but we can have enough criteria for all practical purposes.

People with experience of developing specifications will know what it means to say, "It met the specifications but didn't meet the objective." It can happen. However, the better the quality and the more numerous the items in the specifications, the better the chance of meeting the objectives. It is a tradeoff with what we are able and willing to measure. If we do enough, there's a high probability that what meets the specs will also meet the objective and that the rest will take care of itself. The point is this: *to get criteria, you take each objective in turn and find as many measures as you can which will reflect some degree of achievement of this objective.* In the "ideal" case, we attempt to get as many measures as possible. We can prune them down later—but it is costly and difficult to add them after the design solution is ready.

What is a Good Criterion?

The ultimate purpose of specifications is to screen out unsatisfactory solutions (see Figure 4.1). Those which pass all of the tests in the specifications should meet all of the objectives set for a satisfactory solution.

To be effective as a screening operation, *a criterion must discriminate;* that is, there must be a positive yes or no answer as to whether the design solution meets each element of the specifications. Otherwise, too much argument will result, or as in our initial example, the specification may be watered down until it is meaningless. How to establish discriminating criteria will be covered later in this chapter.

Should Criteria be Developed Before Solutions?

The argument in favor of establishing design criteria in advance of looking for a design solution is: if we establish what the "ideal" solution is in the beginning, then we will continue to look for design

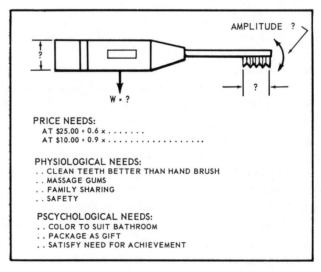

Figure 4.1. Criteria Screen Out Unsatisfactory Solutions

solutions until this "ideal" is found. This procedure reduces the tendency to compromise in seeking a solution. It *spurs* the designer on to create more solutions until he comes close to the ideal. This is the forced iteration that is shown in Figure 4.2.

The Principle of Establishing Criteria

What has been said above can be put in capsule form and stated as a principle.

THE PRINCIPLE OF ESTABLISHING DESIGN CRITERIA

The achievement of design objectives can be controlled by a comprehensive set of discriminating criteria. These should be established before the detailed design solution is sought.

GOING FROM QUALITATIVE TO QUANTITATIVE CRITERIA

If we specify that an engine must have 100 ± 2 horsepower, this is a very objective and discriminating measure. We may need additional references or specifications to define how the measurements are to be made, but in general this is not a problem area. If only all of our criteria could be put down in such numbers! The horsepower scale has some interesting properties. Zero on that scale means no power

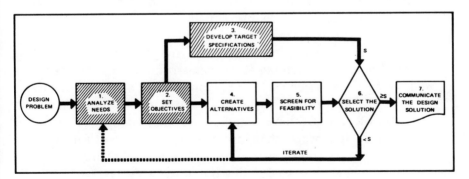

Figure 4.2. Develop Criteria Before Seeking Solutions.

at all. 8 horsepower is twice as much as 4 horsepower. That is why we call it a ratio scale. It is the best kind of scale to use.

Now let us look at another kind of criterion such as "the product shall be esthetically pleasing." Who is going to make this judgment? Should the designer himself make this judgment? Should the marketing manager make it? Should the customers be consulted? Who can really make an *objective judgment* about the esthetics? Well, you can see that as we progress from the designer toward the user, we go in the direction of objectivity. (Albeit an individual user may not be as good a judge as an industrial designer.)

One approach to objectivity is to *refer the design decision to an expert* with an expert judgment on which we can agree. If we give the decision to such a person, the criterion has discriminated between satisfactory and unsatisfactory designs—like a go/no-go gauge.

Another approach is to say that 90 percent of a *statistically sampled panel* of 100 users would agree that the product is esthetically pleasing. This is fairly objective and gets us down to numbers, although we are still using judgment. So as not to be encumbered by too many users around the lab, we may set up a surrogate uninvolved panel of secretaries, technicians, and salespersons who have a discussion and take a vote on whether or not our product is better looking than our competitor's. By such means, we have gone a long way toward objectivity and we do have a discriminating criterion. The design can be categorized as a "pass" or "fail" by the criterion.

There are nominal, ordinal, interval, and ratio scales. In this direction, we progress from pure judgment to objective measures. It is through this sequence of scales that we take a difficult criterion in order to improve its power to discriminate.

Suppose we try to make a specification for a beach. At first we see only beauty, or dullness, or serenity. We are putting it on a *nominal* scale; that is, we are naming it in comparison with other beaches. Could we be more objective? We see sand, rocks, shells, and flotsam. Here are four categories, within each category we could say how many tons of each we have. Already we have made a more useful description of the beach than just to say it is "beautiful" or "serene." Getting the right mix of rocks, sand, shells and flotsam is getting somewhere close to defining the beach we want.

Let us try an ordinal scale now. Which is most important of these nominal categories that we have? To do this we must define a purpose. Once we have a purpose, then we can say which is the most important. If the purpose is sunbathing, then sand comes at the top of the list. If the purpose is souvenir collecting, then the list reads flotsam, shells, rock and sand—in that order of importance. This rank ordering is nothing other than an *ordinal* scale.

Let us try an *interval* scale. Exactly what kind of sand do you want? Is it to be coarse sand with sharp edges, or fine powdery sand? Neither? Well, let us set up an interval scale with 10 grades of sand between the coarse and the fine. You like the stuff near the middle? Will the fourth one do? Right. That's our specification. Notice that on the interval scale there is no zero. Yet interval scales are used all the time for measurement of temperature.

Can we put our beach on a ratio scale? I doubt it. However, parts of it might be put on a ratio scale, parts on an interval scale, and parts on the ordinal and nominal scales.

When designing a beach to a nominal scale, we may prefer to specify the person who makes the judgment as to which named category the beach belongs. An independent inspector would be a good idea. (It's used all the time for acceptance of military products.) This will add objectivity and credibility to the judgments that must be made. So even at the judgment end of our scales, criteria can be made very discriminating by accepting the final word of an independent expert.

If a specification is on a ratio scale, it is very discriminating. If a specification is on a judgmental scale, we can make it discriminating too by referring it to an expert. We can also have a discriminating criterion by having a panel of judges rank it on an ordinal scale between samples A and B. With a little care we can come up with a set of specifications which really pin down the design solution (see Figure 4.3).

TARGET SPECIFICATIONS AND
FINAL SPECIFICATIONS

It is common practice to bring out a document called "design specifications" and call it that whether it is at the beginning or end of the design process. This is frightening. It is no wonder that design engi-

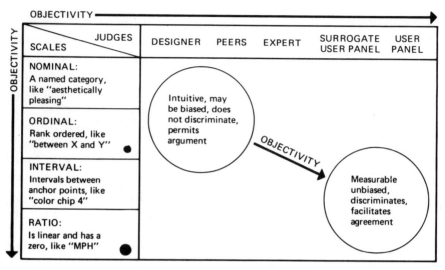

Figure 4.3. Developing Specifications that Pin Down the Design Solution.

neers back off from issuing their design specifications until the last moment. The very word "specifications" has a connotation of being mandatory. Such is not the case when a design begins, and we should properly call them "target specifications." They are targets in the sense that we expect there are to be some tradeoffs made, and some changes in the specifications. In this way, we goad ourselves to achieve the best that we can. We do not lower a target until we have considered all the ramifications.

In the beginning of a design, it is helpful to establish the tradeoff rules. Sometimes this is done by graphs and charts which state what the tradeoff is to be. Examples are: between weight and angular acceleration of an aircraft; between efficiency and cost of a power plant; between water for irrigation and water for electric power, and so forth. In such cases, the target specifications can carry tradeoff data, or even have weights assigned to the various specification elements. These aspects of tradeoff control will be explained in chapter five.

There are some criteria which are not subject to any tradeoff right from the beginning. These I call *"crucial criteria."* They arise from constraints. ***They must be met by the design solution.*** Some crucial criteria are established by the nature of the design itself. For example, there may be a crucial maximum dimension in order to fit a given space; or there may be a maximum torque that may be applied to a shaft because of limitations on other parts of the system. There are also legal constraints which are crucial and which must be met for reasons of safety or for standardization.

APPLICATION TO AN EQUIPMENT DESIGN

Some of the objectives of the previous chapter are developed below into criteria.

Sample Criteria for a Mainly Technical
Equipment Design—A Substation Transformer

OBJECTIVES	CRITERIA OR SPECS
To handle 1000 KVA and transform 13,500 volts to 220 volts at 3 phase 50 Hz.	*When connected up according to test specs xyz, the oil temperature shall not exceed 78° C, at an ambient of 55° C.*
	Test as above, coil temperature not to exceed 200° C.
	After 10,000 hours of operation at full rating, the insulation resistance, as measured by test spec pqr, is not to be less than 50 megohms.
	Etc. for several pages
To be truck transportable from ship to site.	*A lifting ring is to be provided at the top which will not come loose at 3 times the transformer weight.*
	One horizontal clearance dimension to be 10 feet or less when transported. (A crucial criterion).
	Overall height to be between 10 feet and 15 feet.

	Bottom mounting supports to be capable of withstanding a vertical free fall of 6″, or protected to withstand the shock.
	The transformer should be certified as being truck transportable from ship to site by the Office of Transportation in the country of use, or in case of delay, the customer may issue such a certificate.

Comment: It is convenient to reduce the size of technical specifications by referring to other company specs or to industry standardized specs. In this application, the equipment is fairly standardized and the specs are not so much a target to guide the designers as they are mandatory items in a negotiated contract. There is practically no subjectivity involved.

APPLICATION TO A CONSUMER PRODUCT

This application is an extension of the objectives shown in the previous chapter. Note how the subjectivity in the objectives is pinned down by referring the opinion to users or experts.

Criteria Developed from Objectives for an Electric Toothbrush

OBJECTIVES	CRITERIA
1. To be attractive, suitable for sale primarily in the gift market and secondly as a personal purchase.	*1a. Attractiveness of overall design and packaging to be judged better than brands X&Y by more than 75% of a representative consumer panel (ordinal scale).*
	1b. Decorator colors to be same as our regular products (nominal scale).

1c. Package can be displayed on counter area of 3" X 4" (ratio scale).

2. The technical functions are to be at least as good as past "family" models of Brand X.

2a. Technical functions to be judged at least as good as the past "family" model of Brand X by dental consultant, Dr. J. P. (nominal-ordinal scale).

2b. Amplitude to be between $\frac{1}{16}$" and $\frac{1}{8}$" (ratio scale).

2c. Frequency to be 15 ± 5 cps. (ratio scale).

2d. Battery life to be minimum of 50 min. when tested according to standard XYZ (ratio scale).

2e. etc. for other technical aspects such as weight, impact strength, frequency of repair, dimensions . . .

3. To be saleable in US and Canada.

3. Must meet UL & CSA standards for safety (a crucial criterion).

4. The timing objective is that the product be ready for sale to the Christmas trade in the nearest feasible season.

4. The time milestones, backing up from October production are to be:
• mock-up approval—2 months
• tooling release—6 months
• production prototype—10 months
• pilot run—10 months
• production run—13 months
  (October)
(Calendar time is an interval scale, elapsed time is a ratio scale.)

5. The selling price is to be not more than 10% of the present utility models.

5. The selling price is to be between $12.50 and $17.50, depending on the features offered, for a production run of 100,000 units (money is a ratio scale).

Comments: Note the use of two means to get a discriminating and credible decision on the first two criteria by use of (1) an unbiased reference panel and (2) a technical expert other than from the designers themselves.

Note also that an objective, particularly a technical objective, can lead to many, many criteria. The example we have is largely representative of "target" specifications for an upstream design activity. At the decision point of handing over the design to production, the specifications would be much more extensive.

APPLICATION TO A LARGE-SCALE PROJECT

When millions are spent on a capital project, much effort goes into developing specifications. These may be flexible target specs at the beginning, but they will be very rigid at the point where the construction contract is let. The designers cannot afford to experiment, or work it out as they go, when millions of dollars are involved. It is worth the effort of developing specs that pin it down.

Example of a Large-Scale System—a Hydro-electric Dam for Latin America

OBJECTIVE	CRITERIA
1. To produce 400 megawatts of electric power at start.	1. Produces 400 ± 50 megawatt at start. Weight, .4
2. To produce 50,000 inch-acres of irrigation water.	2a. Produces 50,000 ± 20% inch-acres of irrigation water. Weight, .2
.	.
.	.
.	.
.	.
Q. To provide an object of national pride.	Q-1. To provide an object of national pride which is agreeable to 75% of the Cultural Commission. Weight, .05

Comments: In this example, the objectives themselves were quite measurable because most users' needs had been transformed to electric power usage based on historical data of other installations. Nevertheless, to reduce the argument over what is satisfactory, limits have been put on the major parameters. In the beginning of such a design project, the site and size of the dam have a big bearing on the outcome.

Besides limits, we have given relative weights to the criteria to give guidance to those who must select a site and make a subjective as well as a technical tradeoff. The target specifications contracting the design would be volumes, and the final specs for construction would be much more voluminous, including many specification drawings and standards.

APPLICATION TO BUILDING CONSTRUCTION

The development of target specs from objectives

Objective P1. *A cafeteria suitable for timely serving of hot meals to 200 persons per staggered meal break.*

- *Criterion 1a. If a group of 17 persons enter the cafeteria every 10 minutes over a 2 hour period, the average waiting time per group for serving a full meal will not exceed 10 minutes.*
- *Criterion 1b. At a flow of 17 persons in 5 minutes, there should be table space available, assuming that no person stays there for more than 40 minutes of the 45 minute break.*

Objective P2. *Fast service for beverages and light snacks.*

- *Criterion 2a. Waiting time for hot beverage and snack not to exceed 5 minutes for 15 out of 17 persons entering the cafeteria every 10 minutes over a 2 hour period.*
- *Criterion 2b. Three kinds of hot beverages and six kinds of cold beverage should be available for fast service.*
- *Criterion 2c. Snack service to be conveniently located for access to the TV room and games room.*
- *etc. There will be other specifications relating to the use and structural integrity of the building. Many of these will be references to building codes and architectural standards. In all the target specs could be 100 times what is here in this example.*

CASE HISTORY OF A BEVERAGE CAN OPENER

Not long ago, before tab-openers were developed, beverage cans were opened by devices which punched two large holes into the top. Because the devices on the market could not be used conveniently by children, a need was perceived. A design team looked into the matter and came up with the following preliminary target specifications.

SUBOBJECTIVES	CRITERIA
1. Easily operated by children.	1a. Grip pressure – 10 lbs. 1b. Lever pressure – 18 lbs.
2. Easily portable.	2a. Maximum length – 7 3/8" 2b. Weight – 5 oz. 2c. Width – 1 5/8" 2d. Height – 1 3/4"
3. Rustproof.	3. Should pass test of 2 years simulated normal use, using industry standard test XYZ.
4. Additional subobjectives were: must be sanitary, compact, safe in operation, attractive, reliable, reproducible, easy to clean, operable with one hand, have a spray guard, no sharp projections.	4. These would be defined at a later stage in the design.

Comments: You can see that while some numbers have been assigned to the first objective, whether or not the device is easily operated by children would not be fully settled by the application of these two criteria only. There probably should be some kind of test conducted with a number of children, or an expert should be consulted.

This is an example of a systematic attempt to define criteria before the product is actually designed. Admittedly, a beverage can opener may not seem to justify a lot of effort. On the other hand, how many gadgets reach the market and are entirely unsuccessful? Plenty. Either they don't function well or they don't meet a need. In this case history, a second analysis of needs led to the finding that the device would not be needed in 2 years because of the advent of the tab-opening cans.

DESIGN GUIDELINES

<table>
<tr><td colspan="2">GUIDELINES FOR DEVELOPING TARGET SPECIFICATIONS</td></tr>
<tr><td>Do

Do develop target specs while the analysis of needs and objectives is fresh in your mind.</td><td>Don't

Don't wait until the design is finished to write up the specs. They will suit what you have done but you would be safer to buy a ticket to Timbuctu.</td></tr>
<tr><td>*Do make sure that the objectives are well covered by criteria to measure their achievement.*</td><td>*Don't mistake the directional sign-posts for measurements.*</td></tr>
<tr><td>*Do change intangible factors into discriminating criteria.*</td><td>*Don't believe for a moment that you can win out on the intangibles by a snow job.*</td></tr>
<tr><td>*Do improve the objectivity of your target specs.*</td><td>*Don't allow the design effort to become a free-for-all about what is right.*</td></tr>
</table>

PIN DOWN WHAT YOU HAVE LEARNED

Without a systematic approach to design, it is easy for you to get carried away with the fun of creating something new. You need not give much attention to being specific about what is needed or when it is needed. You can just keep busy doing your own thing—but only up to a point. We have a little problem here. Making plans and target specifications which define what we are trying to achieve is not nearly as much fun as working on the design itself. That's the rub. Reasoning and logic are less satisfying than design.

In the long run though, it is better that we discipline ourselves to be systematic rather than having it done by others. The way to discipline yourself is to practice the development of target specifications by the following exercises on your practice project or on a real design project of your own.

I have told you ways to get discriminating criteria for the intangible factors of an engineered design. Can you really do it? Not.unless you practice. Why do I remind you of this? Simply because so many engineers have told me that while the technical parts of the specifications are easily defined, it is the intangible ones that give them difficulty. So go to work with the following exercises. You can then be sure that you can obtain specifications from objectives, use several scales of measurement, and be able to apply the test for good criteria. Your specifications will be meaningful and you will be able to reduce the disputes you may have over what really meets the objectives.

**FOLLOW-UP EXERCISES TO HELP YOU
BECOME A PRO**

**Questions to Help You Consolidate Your
Understanding of This Chapter**

1. What is the main attribute of a really good criterion or item of a specification?
2. What is the difference between target specifications and final specifications?
3. Is an ordinal scale more objective than a nominal scale?
4. For those intangible factors that can't be quantified, how do you go about obtaining discriminating criteria which have objectivity?
5. Take the criteria for the electric toothbrush and mark their numbers in the appropriate place on the chart of objectivity of Figure 4.3. Can you now make them more objective?

6. What is the advantage of giving relative weights to the technical parameters of a specification, such as was done for the hydro-electric dam?

Questions on the Case History

1. Are the criteria for the first three objectives truly workable? In other words, do they discriminate between satisfactory and unsatisfactory solutions?
2. What modifications would you make to the criteria if the can opener is to be used for opening 25 gallon, nonreusable bulk chemical containers?
3. In what other ways can you pin down the satisfaction of the subobjective, "easily operable by children" without using the ratio scales?

Assignments that Help You Become a Pro

1. For your practice project or for your real project, take the objectives you developed in the prior chapter and use them to develop target specifications. Pay special attention to quantifying the objectives for those intangible factors.
2. Take an engineered design you are familiar with and find a suitable model under the applications to equipment, product or large-scale capital projects. You may already have an existing set of specifications for one of these. Either develop a comparable set of specifications for the engineered design or go over the existing specifications and try to improve them. Develop examples of improving the objectivity of your criteria by using the technique of more rigorous scales and also by more objective opinion.
3. If you have contact with the working world, ask your colleagues for examples of specification items that caused them difficulty. Take these and improve them by using the concepts of this chapter. Then, resubmit them to these persons for their opinion. By this means, you will improve your skills.

5 Keep the Project On Target By Controlling the Tradeoffs

KEY IDEA: FOR A PROJECT TO REACH ITS TARGET FOR PER-
FORMANCE, COST, AND TIME, YOU NEED PLANNED CONTROL
FOR THE MAJOR AND MINOR TRADEOFFS.

THE VEHICLE THAT WORKED TOO WELL

A brilliant mechanical engineer had an idea for a commercial type snowmobile. It would serve the need for winter travel over snow-covered power line and pipeline routes where tools and equipment would be carried along for line repairs.

He was very capable and had designed some outstanding vehicles for his former employers. He managed to raise capital through his track record, and also by mortgaging everything he had. He went into business for himself to produce commercial snowmobiles.

He built one prototype and tested it, then another and tested it, and so on. His business philosophy was to have a high-performance, reliable vehicle that could sell against all competition for years to come.

Here is an example of one way in which he applied his philosophy. The engine required was much heavier than any of those available for recreational snowmobiles. He shopped around till he found a good engine. However, he did not like the power curve and decided to make some improvements with it. This involved complete redesign and testing of the muffler system because it was a two-cycle engine. The muffler design, by the way, required a lot of skill and a lot of iterative design as there was no formula that told him exactly what to do.

Because of the engine redesign, more prototypes and more tests were necessary. An extra year was added to the original target for completion of the design. Consequently, his financial backers were less than sympathetic when they realized that this engineer was more concerned with the perfection of his toy than with getting them the profits he had promised. His expenditures went up. He needed to buy more engines, he needed more money to pay the mechanics who worked for him. In the long run, he got the design ready, but only by using up all of his available capital, plus a lot of extra time.

He managed to save his house by selling his design to someone else, and another person made the profit from his enterprise. His friends sympathized with him. If only he hadn't tried so hard for technical sophistication! There were many engines that would have performed reasonably well and served the need, but he did not know how to control the major tradeoff of performance with time and cost, and keep the project on target.

WHAT YOU WILL GAIN FROM THIS CHAPTER

Post-mortems of many design projects that missed their targets for performance, cost, and time have shown that there was a lack of tradeoff control. This chapter will show you what to do right from the beginning to control the forces that would otherwise give you an unbalanced result.

Objectives

1. *You should learn and understand the essential interrelationships of performance, cost, and time on a design project.*
2. *You should learn the difference between major and minor tradeoffs.*
3. *You should learn how to mediate the tradeoffs by technical considerations, by cost considerations, or by trading satisfactions.*

Benefits

1. *You will be able to use tradeoff control to effectively meet design project targets.*
2. *You will be able to do a tradeoff analysis for deciding on major tradeoffs.*
3. *You will be able to demonstrate your ability to use the sophisticated techniques of tradeoff control.*

Tradeoff control is not one of the seven steps within a design phase. It is a process or way of thinking that you apply throughout the design process from beginning to end. In the beginning you establish guidelines, you exercise control throughout, and in concluding, you shape the overall results.

In the beginning of this chapter, you were introduced to the case of the engineer who lost his own business because he could not control major tradeoffs. Now, we will generalize from this case and others up to the "principle of the tradeoffs." Under the applications of the principle, you will find an example of a technical tradeoff between electric power and irrigation water for a hydroelectric dam; tradeoff of efficiency and power in the choice of an engine for an equipment design; and the assessment of the overall percent satisfaction for the technical and nontechnical tradeoffs in a consumer product. You will then have learned what you should do about the major tradeoffs. You will have three distinctly different tools to use for different situations on your design projects. Then we go into the matter of controlling the design team on a project by using review meetings to control major tradeoffs and by using guidelines for the minor tradeoffs. All of these things are summed up in Guidelines for Controlling Tradeoffs.

For those who want to go further and develop their skills in controlling tradeoffs, there are exercises at the end of the chapter.

THE PRINCIPLE OF TIME TRADEOFF

The attributes of a design project can be classified into:

Performance attributes, P, such as the specifications.

Time attributes, T, such as the schedule for design completion, production, and use.

Cost attributes, C, such as the cost of the human resources needed, the cost of the materials, the cost of making and distributing, and the profit if it exists.

For example, a crash program to get a passenger vehicle on the market in time to meet competitive efforts might require the sacrifice of some efficiency (performance) to complete the project in less time and keep the cost essentially the same. Thus, performance, P, would be traded off for time, T. You might say, in this instance, given that the efficiency was above the threshold level, that T was preferred over P. We express this as:

$$T > P \Big|_{C \,=\, \text{const.}}$$

If the cost is frozen, then the target time, T, can be reduced by reducing the performance level, P. This would be the guideline for a major tradeoff.

Similarly, we could imagine a case where circumstances caused the performance target to be revised upwards, such as a demand for a higher efficiency engine because the price of fuel had gone up immensely. Then the design authority may be prepared to put more resources into the project and still complete it in the same time. They can improve the efficiency (performance, P) at a small increase in the final cost of the product. We can express this major tradeoff guideline as:

$$P > C \Big|_{T \,=\, \text{const.}}$$

If the time is a fixed deadline, then an increase in the performance level, P, can be obtained by committing more resources, C; given the P, T, and C are all above threshold values.

This fundamental relationship of P, T, and C is understood by pros in the design game. It is a fact of life: P, T, and C, become related in a project or task.

THE PRINCIPLE OF THE TRADEOFF.

The Performance Level Achieved by a Design is a Tradeoff With Both the Cost of the Resources Used and the Time Taken.

If there is to be a higher level of performance achieved, then there must be either:
  1. *more cost for materials and resources used, or*
  2. *more time taken, or*
  3. *a tradeoff with both cost and time.*

AN APPLICATION OF CONTROLLING THE TECHNICAL TRADEOFF IN A HYDROELECTRIC DAM

Two objects cannot occupy the same space at the same time. Therefore, there are all kinds of dimensional tradeoffs that are necessary. The trunk of a car can be larger if the seating space or gasoline tank size is reduced. The steering wheel can be larger if the driver space is reduced.

Take the example of water, stored behind a dam. If released through turbines, it can generate electric power. If the water is taken from the top of the dam and used for irrigation, then it cannot be used for power. Water for irrigation becomes a technical tradeoff with electric power because the potential energy of the water is lost when it seeps down through the ground or evaporates from the surface.

If you have a stereo power amplifier, you turn the volume up high and get more power output, but you generally get more distortion too. Transistors and power supplies are obliged to obey the physical laws of nature. Consequently, more power means more distortion. A technical tradeoff cannot be avoided.

A technical tradeoff relationship can generally be expressed by a formula or a curve. The question that must be answered is: Where do we want to be when a tradeoff is made? Take, for example, the tradeoff shown in Figure 5.1, between electric power and irrigation water. This could be developed from the topographical features of various dam sites. If we have a minimum figures of electric power and irrigation water, we have a limited region in which a tradeoff can be made. A technical tradeoff is possible but some guidance is still needed. Before the tradeoff curves are developed, one can specify a general preference; but after they are developed, then decisions must be made as to what the final tradeoff will be. If this is a major tradeoff, it will be dealt with by the planners at the review meetings.

The point about technical tradeoffs is that the tradeoff relationship is determined by the nature of the design itself, but the operating point is subject to some judgment. When judgments come into play, we may be able to revert to cost or time in order to mediate the tradeoff.

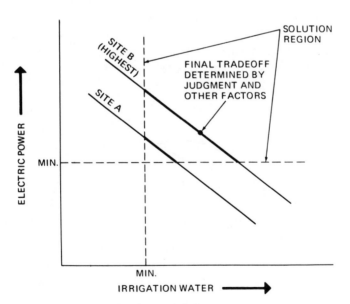

Figure 5.1. Technical Tradeoff Relationships.

AN APPLICATION TO AN EQUIPMENT DESIGN
WHERE THE COST MEDIATES
A TECHNICAL TRADEOFF

Suppose, for example, that we are designing an all-terrain vehicle, such as the snowmobile mentioned earlier, or some other special vehicle. If we do not design the engine from the ground up, then we must choose among those which are available. Suppose that we have three engines to choose from and we are concerned about the trade-off of horsepower and efficiency. These are not related in a technical sense (except by some stretch of the imagination) and we must revert to a third parameter, such as cost, in order to make the tradeoff.

The available horsepower and efficiency will be determined by the speed of operation. Given an application where the speed is fixed, we can choose from engines with various efficiencies, but the cost will vary. Essentially, we want the most efficient engine that we can afford. But what can we afford?

A gasoline engine has a peak efficiency at one particular speed. Below this, the frictional losses are significant and above it, the combustion losses are significant. (To some this will be an over-simplification, but the example will serve to make the point.)

Suppose there is a choice of three designs: A, B, or C, such as shown in Figure 5.2. In this figure the speed has been transformed

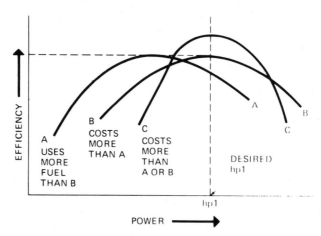

Figure 5.2. Efficiency-hp Curves of Engines A, B, and C.

to horsepower by assuming contant torque. Machine A has its peak efficiency at a speed below the desired out of hp_1. Machine B has its peak efficiency at hp_1, but is larger and has an incremental amortized cost of \$100 more per month. Let the incremental fuel costs of machine A at hp_1 be \$50 per month. In other words, if cost is the main consideration, then A is preferred on a life-cycle cost basis.

APPLICATION TO A CONSUMER PRODUCT
WHERE THE TRADEOFF IS IN SATISFACTIONS

Judgmental Tradeoffs: Bartering

A judgmental tradeoff depends mostly on circumstances and not upon the object being designed. We use our judgment to get the best deal we can under the circumstances—a bartering process.

A judgmental tradeoff is not determined by the physical laws of nature as a technical tradeoff might be. For example, while an automobile trunk space is a tradeoff with seating space or tank size in the technical sense, what is actually done about it is a matter of judgment. For a traveling salesperson, the trunk space is valued more than the other spaces and it would be favored in a judgmental tradeoff.

In another circumstance, such as cars for the taxi market, you simply must have leg room and luggage space, and given the need for short overall length, it may be necessary to cut down on the size of the gas tank.

A judgmental tradeoff is entirely dependent upon the value system of the decision maker. It is not a linear function. Consider an automobile which is to be efficient and also to have good appearance. The decision maker may have satisfaction functions such as those shown in Figure 5.3. Above threshold values, the present satisfaction increases rapidly until saturation is reached.

The S-curve is similar to most physiological and psychological scales, such as the scale for sound level, also shown in Figure 5.3. However, since sensation varies as the log of the stimulus, the S-curve can be transformed to relatively linear scales by taking the logarithm of the parameter scale. Note that the linear scale is not extended very

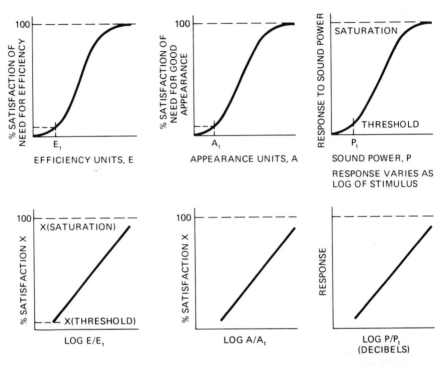

Figure 5.3. Converting Parameter Scales to a Common Percent Satisfaction Scale.

far into the regions below threshold or close to saturation. These are special cases. For example, if the appearance is below a certain threshold value, it simply cannot be traded off for efficiency.

It is satisfaction that can be weighted and added. We transform the parameter scales to percent satisfaction scales which are relatively linear. We can now approximate the valued tradeoffs with a linear tradeoff function:

$$\text{The Overall \% Satisfaction} = w_i X_i$$

Where

X_i is the % Satisfaction of the (i)th need (objective or criterion)

Where

w_i is the weight assigned to the (i)th need (objective or criterion)

Subject to conditions:

$$\sum w_i = 1.0$$

and

$$X_{\text{(threshold)}} < X_i < X_{\text{(saturation)}}$$

As a tradeoff control, the linear tradeoff function is a way to measure the true value of improvements or losses in performance. We can show that this tradeoff function can be extended to include the cost and time aspects of the design.

Suppose that during the design phase of the electric toothbrush there was a proposal to have some adjustment holes in the case so that the factory could adjust the mechanism after it was assembled. When needing repair, it could be taken apart by screws reached through the holes. Customarily, industrial designers do not like access holes showing on the appearance side of a consumer product. Even if they put plugs in them they spoil the smooth appearance. (By the way, I have gone through this argument between the industrial designer and my mechanical engineers. The engineers wanted an access hole on the front panel but could not convince the industrial designer that it was necessary.) You may recall that the esthetics are important because the product is to be a gift item. The technical functions needed to be only as good as "brand X."

To solve this problem, we must construct part of a decision matrix. (This topic will be covered in more detail in chapters eight and ten.) Refer to Figure 5.4. The first column is entitled "esthetics." We see that the no-holes design option rates 95 percent satisfactory in this column. The rating is only 70 percent for the holes-in-the-cover option. The technical functions rate correspondingly 80 percent and 90 percent. These are judgmental values arrived at by opinions of objective persons. Clearly, the gain of 10 percent in technical functions is not as much as the loss of 25 percent in satisfaction from the esthetics column. To make matters worse, the esthetics have a weight .3 and technical function a weight .2 which tips the decision even more in favor of the no-holes option. Imagine, if you will, a utility model of this product where the esthetics are given a weight of only .1 and the technical functions a weight of .3. Then the difference of technical functions would be three times as important as the difference on the esthetics. That would slightly favor the holes-in-the-cover option.

For the purpose of this example, we have assumed that the unit cost and the completion time were not affected by this tradeoff.

Note also that the satisfactions in all cases were well above the threshold values and in the case of esthetics were approaching saturation. It is unlikely that any proposals to further improve the esthetics would buy enough satisfaction to warrant any tradeoff with cost, time, or technical functions.

USE REVIEW MEETINGS TO CONTROL THE MAJOR TRADEOFFS

Control of performance, cost, and time is exercised through review meetings held at major milestones and at the end of the major design phases. Tradeoff control is also exercised there. Most of the major tradeoffs come to the attention of top management before or during these meetings because they require management decisions. A major tradeoff gets attention because it may need more funding, may require substantially more time, and/or may require negotiation of a new contract. There is also the possibility that the physical benefits may be affected.

If, at these review meetings, we bring our tradeoff data in one of the forms shown above, it will be possible to make sensible decisions

OBJECTIVES / WEIGHTS / ALTERNATIVES	ESTHETICS AND GIFT ITEM	TECHNICAL FUNCTIONS AND SERVICABILITY	READY ON TIME	COST	ETC.
	.3	.2			
NO-HOLE	95%	80%	–	–	–
HOLES-IN-THE-COVER	70%	90%	–	–	–

Figure 5.4. A Tradeoff of Satisfactions.

without counter-productive conflict. Whether we like it or not, there will be social pressures within the organizations by those jurisdictions representing performance, cost, and time. We should be sure that the outcome is not decided by strength of personality. The facts should make their own case.

DON'T FORGET THE MINOR TRADEOFFS

Investigations of the reasons for overrun in time and cost on a number of projects have turned up the interesting point that many of these could be traced to cumulative design changes. The small design changes which do not catch the attention of management, when added up, constitute a formidable pressure on time and cost. (The incremental performance, ΔP, causes an incremental change of ΔC or ΔT.) It is difficult for the people in charge of the design project to control all of the minor design changes that are made at the lower levels in a large organization. The draftsman calls for six rivets instead of four, and this is not noticed. An engineer specifies a higher quality component because he didn't get the other samples in time. Little by little, the performance can "drift" upwards and the cost and time will follow.

The "drift effect" makes it difficult to keep the performance, cost, and time close to the target because of irreversible design changes, cost increases, and time taken. *The way to cut the tendency to drift is to make the minor tradeoff guidelines explicit.*

At any point in a design process, it is possible to state whether performance, time, or cost dominates the minor tradeoffs. This does not give one license to make any change whatsoever, but provided that performance, cost, and time, are within reasonable limits, there is elasticity in these. Control the minor tradeoffs by a statement similar to the one that follows. (This is also described in chapter three.)

Minor Tradeoff Guideline

Given that performance, cost, and time are in the satisfactory region, then, at this point of the design process, the incremental performance, cost, or time (only one) dominates over the other two.

This should be updated as the design progresses. Normally, the performance will dominate in the beginning of a design project, and the time will usually dominate near the end.

In a design project with very tight cost control, we may control the minor tradeoffs by a Minor Tradeoff Guideline that states:

Given that P, T, and C are in the satisfactory region, the preferred minor tradeoff is:

$$\Delta C > \Delta P, \Delta T$$

In plain language, what this means is: Keep the cost under tight control! Don't let it increase for minor design enhancements! Don't let it increase to save time—but if you must make a major tradeoff beyond the satisfactory region, let the boss know!

DESIGN GUIDELINES

GUIDELINES FOR TRADEOFF CONTROL	
Do	Don't
Do define the tradeoffs as much as is practicable early in the design process.	*Don't postpone changes, and then get into hassles over tradeoffs when it is expensive to change the targets.*
Do try to establish major tradeoffs by preparing data on the tradeoff relationships and deciding on the operating point at a review meeting held at a major milestone.	*Don't let intuition carry the day when facts could be presented.*
Do control the minor tradeoffs by making known what is wanted when a tradeoff must be made.	*Don't leave the marginal tradeoffs to personal whims or pressure tactics.*

A CASE HISTORY OF THE THREE-WAY
TUG-OF-WAR

Let me tell you about an aggressive and successful manufacturer I knew. His company did its own engineering on its products which were quite successful in the marketplace. The president and owner of the company was an engineer himself and dedicated to high quality products. He believed in a fair deal for the customer and for the manufacturer. He hired a stable of engineers and technicians who believed in this philosophy. They ran the company, and while the products didn't sell very fast, at least they stayed in business. Somehow, they survived the first rash of price and quality cutting that threatened the industry. At the point where the customers were looking for quality instead of price, they were ready to turn on the production. This they did and the company grew very rapidly with products of high technical sophistication. Performance and quality were household words. ($\Delta P > \Delta C, \Delta T$)

In a few years, the cycle reversed and costs rose to paramount importance. This company had grown to the point where they needed to have volume sales to sustain the level of business to which they had become accustomed. They cut prices without cutting back on quality—and nearly went under.

The president, who was going into semiretirement, reacted sensibly and hired reliable financial controllers. They instituted budget planning and expense recording. All jobs were properly cost-estimated and care was taken to see that expenses did not exceed the supply of cash generated by sales. The financial people did very well and were soon put in control of the company. Eventually, they moved in on the engineering department and told them to cut their staff and to cut the manfactured cost of the products as well.

The engineers had been trained in the philosophy of high quality products for years, and they weren't about to change because of a few financial people telling them what to do. Instead of dropping the performance, they did the only thing possible—they simply took longer to complete the designs. ($\Delta C, \Delta P > \Delta T$ was what they were doing). The controllers thought they had been successful as products started coming down with a lower engineering cost. Unhappily, however, the company began to lag behind the needs of the market.

The timing was just not right. Products at the right price, which are out of season or out of style with competition, are not hot sale items.

Now it was the marketing manager's turn to make his move. If anyone knew of the importance of timing in the marketplace, it was the marketing manager. He, fortunately, had the ear of the president. A quick shuffle at the top put the marketing manager in charge. He brought in a whole new crew of executives and managers who were determined to get the products out on time, and not surprisingly, they succeeded. The financial people still had all their rules and regulations and they exercised control (ΔT, $\Delta C > \Delta P$). There was only one way that design could move and that was to lower performance level. The reliability and the performance soon fell. By this time, the quality-minded engineers and technicians had moved onto other companies and were giving their old company a hard time by designing competitive products without undue pressures on cost and time.

I would like to be able to tell you that this company learned a lesson and now keeps performance, cost, and time in control, but that is not the case. Like many companies I have watched, they go through cycles of pressure on one or both of these. They never seem to get the idea that they are all interrelated and that they can only have guided control.

When the performance, cost, and time characteristics of a design result in a 3-way tug-of-war in an organization, as it frequently does, then the tradeoff goes in the direction of the group with the strongest ego, the strongest personality, and/or the strongest clout. Tradeoff control cannot be left to these kinds of circumstances. It can be planned, it can be executed, and the results can be gratifying.

A CASE HISTORY OF CONTROLLING THE
MINOR TRADEOFFS FOR A TV SET

Long before I learned about proper tradeoff control, I had an experience which I remember to this day as an example of its application. I was in charge of engineering design for a medium-sized TV manufacturer. We were well aware of the importance of performance, reliability, producibility, and cost. However, with each design we were faced with dozens of decisions between these attributes. For example, some additional performance usually involved some marginal and additional costs. How much tradeoff was warranted?

John S., the engineer in charge of our first color TV set design, came to me and asked for tradeoff guidance. He wanted to know which was the most important:

Performance

Serviceability

Producibility

Cost

I took this matter up to the product planning committee and the answer I got was that these were all important—back to square one. At the insistence of my design engineer, I went back again. This time we made a joint proposal on what we thought the rank ordering should be. We argued that since the company was entering a new field, we should establish our reputation on the basis of high performance and serviceability. The selling price was mainly established by other manufacturers. Since our product was going to be much more expensive than that of our competition and there was little we could do about it, we thought the cost should be last in importance. Moreover, since the quantities to be produced in the factory were small, we thought that producibility was not very important. Our proposal shook them up a bit and they agreed with our ranking more or less, but said that they should be "ranked in close order." Whatever was meant by that we did not know at the time, but I am sure it pertained to the fact that these things had to be in the satisfactory region before we could exercise these guidelines for the marginal tradeoffs.

As a result of guidelines which were as enumerated above, the design engineer decided, on his own, to adopt a superior-type color picture tube which increased the cost by a significant amount, but was still in the satisfactory region. Because this picture tube resulted in 30 percent more brightness, we had a selling feature over our competition. The marketing department was delighted with the results. Another minor tradeoff that was made was to put an access door on the front of the TV set so that the serviceman could make the color adjustments from the front. This cost a little bit more but it ingratiated us with the servicemen who tended to recommend a TV set that they were able to service easily.

I don't know of all the minor tradeoffs that were made. These, as they always are, were in the hands of the design engineers who handled those details. All I know is that the end result was very good and tradeoff control worked for us.

WHAT YOU LEARN IS A TRADEOFF TOO

It is true that the more effort you put into learning the ideas in this book, the more you will learn, but the tradeoff is with time that you can put into other things. What about that? If you read this chapter, you may retain about 10 percent of it. If you apply the ideas to a real problem you will probably remember them forever. In other words, application of this chapter to real world problems is probably 10 times as effective as reading about it. In between is the increased effectiveness you would get from doing practice projects. Yes, it is a tradeoff, but by reading only, you are barely above the threshold value. Anything more you do will be fairly efficient learning and you will probably not get into saturation until you have done about a 100 hours of follow-up.

Major tradeoffs are generally not a problem with engineered designs. These generally require management's okay to change the basic targets for performance, cost, or time. Certainly, on the big projects this is true because the extra money is not available without some discussion. Smaller size design projects can have major tradeoffs made without consultation of the users or sponsors and, therefore, major tradeoff guidelines would be useful in these cases.

Where tradeoff guidelines are most useful is for the *minor* tradeoffs. These tradeoffs are made at the working level and can go any way at all unless under some form of control, such as by guidelines for the minor tradeoffs.

In this chapter, you have learned how to do a tradeoff analysis for recommending or making a major tradeoff. You should also be well aware of the importance of making known the guidelines for the minor tradeoffs. In the remainder of this chapter are questions to refresh your memory of the concepts by recall, and also some think-type questions about the case histories. Then, there are specific assignments on your practice project (or a real project) to help you become top notch. Actually, not many people know what they should know about tradeoff control. Learn it well and you will make your mark.

FOLLOW-UP EXERCISES

Questions for Reviewing Ideas in this Chapter

1. We don't want to give design engineers license to make unreasonable tradeoffs. We give them guidelines but we also include a statement that limits their actions. What is this qualifying statement?
2. What would be your approach to making a decision on a purely technical tradeoff?
3. What would be your approach to making a tradeoff between tangible and intangible factors in a design?

Questions on the Case Histories

1. Refer to the case history on the Three-way Tug-of-War. What could this organization have done to reduce the extremes of the emphasis caused by their changes in management philosphy?
2. In the case of the color TV design case history, do you recognize an important design factor that was not listed in the tradeoff guidelines?

Assignments to Help You Become a Pro on Tradeoffs

1. For your practice project or a real project, find or imagine the three kinds of tradeoffs enumerated in this chapter. They were: (a) a purely technical tradeoff, (b) a cost-mediated tradeoff, and (c) a satisfaction tradeoff. Work them through to demon-

strate that you understand the technique. If you cannot find a satisfactory one on your project, simulate one by an engineered design similar to one of the examples in this chapter. A good chance for tradeoff analysis and control doesn't often come so you may forget about these techniques if you don't practice them on something soon.

2. If you are a practitioner, try and recall a case history in which there was inadequate tradeoff control. State what you would have done to exercise proper tradeoff control.

6 How To Get Good Ideas

KEY IDEA: YOU KNOW YOU HAVE A GOOD IDEA WHEN YOU
HAVE COMPARED IT WITH A WIDE RANGE OF ALTERNATIVES.

THERE IS ALWAYS ROOM FOR IMPROVEMENT

When I was in charge of an engineering design group, we had one
television design we loved so much that we named it the "Mother
Chassis." It was so well designed that it became the basic unit for
most of our television products. An unusual group effort had gone
into this design and we felt that it had the most performance for
the least cost. When asked to reduce its costs even further, we felt
that this could not be done. The design engineers had put their best
efforts into maximizing the ratio of performance to cost when it
was designed in the first place.

Although our first reaction was "It can't be done," it seemed that
everyone had a few ideas that could be tried out. There were three
reasons why there could be reductions in its unit cost. They were:

1. Years had passed since the original was designed and there were
   changes in the prices of components.
2. There were new opportunities because of the high volume that
   production had reached on the "Mother Chassis."

105

3. There were technological changes that could be incorporated into the design.
4. The customers' needs had changed in some areas.

One of our clever young engineers was assigned to the cost reduction program. Because he had not been involved in the original design, it was his job to solicit ideas from other persons and to test them out. He went to all the electrical and mechanical design engineers in our group, broadened his quest to include our "rivals" who designed other kinds of products in the company, called on vendors to give ideas, and conducted a number of group brainstorms. He collected over a hundred ideas. He tested them out and many proved to be effective. For example, a vendor who sold us radio frequency transformers within a metal shield can, suggested that we might eliminate the cans by some minor redesign. This idea was developed further by a brainstorming team and eventually cut the cost of these components in half. (The vendor seems to be cutting his own throat in this example, but since he was purchasing the cans anyway, the vendor's profit on the new coils was as good as before.)

Every idea that was adopted nibbled away at the cost of manufacturing. The end result was a three percent reduction in the cost for materials and labor. Since the profit margin on this consumer product was not large, this corresponded to an effective 15 percent increase in the net profit on that product. After that we never again said, "It can't be done."

PUT YOURENERGY WHERE IT COUNTS

As engineers, we have the reputation of being action-oriented people. When given a design assignment, we jump right in and expect to come up with a result. Jack E. was a bright young engineer when I knew him. He never lacked for ideas. When he got one, he would immediately build a prototype to see if it would work. If it didn't work, he would sit down at his desk and think up another idea. Then he would go to his workbench and try it out.

After some training in better design methods, he found that he could save a lot of time by simply asking himself the question:

What are the alternatives? When Jack got an idea, instead of jumping up and getting to work on the bench, he would wait until he had a number of other alternatives. He would then think them through and choose the best of these, *then* he would go into action. This little change doubled his effectiveness. He told me later, "It's the results that count, and not how hard or fast you work."

WHY WE SHOULD NOT GO
WITH THE FIRST "ANSWER"

If we consider solutions to problems such as: 2 times 2, or 2A plus 2A, we find that we have only one solution. Hence, our first "answer" is *the* answer. Most of our school training leads us to tackle problems with the object of finding *the one and only solution*. But consider the following problems:

- What is the correct mix of cement, sand and crushed stone for concrete?
- What is the correct L/C ratio for a tuned anode electronic amplifier?
- What is the correct way to teach problem solving?

A search for the one "correct" solution to each of these problems is fruitless. They are open-ended. There are many "answers."

Consider this problem: How can an electrical connection be made between a 20-gauge cadmium plated steel chassis and a 22-gauge hookup wire? Here are some of the "solutions."

1. Solder directly
2. Solder to riveted terminal
3. Use machine screw
4. Use self-tapping screw
5. Use spot welding
6. Use capacitor discharge welding
7. Use a laser beam weld

The first is the conventional solution and is the best in many cases. However, if you are cramped for chassis space or working on a crowded prototype, the sixth may be better than the first. The "correct solution" depends on the circumstances.

WHAT THIS CHAPTER IS ALL ABOUT

When you use my systematic design procedure, you will come to a point where you will systematically create alternatives as indicated in step 4. You should be able to create a wide range of design alternatives before working further on evaluating their feasibility and selecting the design solution. This chapter will help you to enhance your creative output and at the same time show you how to put your energy where it counts.

Objectives

1. *You should learn how to tap the world's storehouse of experience.*
2. *You should know about five mind-expanding ways to get ideas.*

Benefits

1. *You will always be able to develop a long list of conventional alternatives.*
2. *When you need new ideas, you will know how to get them.*
3. *You will be able to nurture your creative talent and get progressively better at getting new ideas.*

First, you will be shown four ways to tap the world's storehouse of experience. Then you will learn five ways to expand your mind's capacity to get ideas. These are the SCAMPER Sessions with Deferred Judgment; Incubation and Illumination; Morphological Analysis; Personal Analogy; and the Idea Trigger Session. Applications to an equipment design, a consumer product and a large-scale capital project will give you models for generating ideas on your own designs. A real case history of equipment design shows how the team came up with really creative ideas after they had pressed through to the seventeenth. All of what you have learned in this chapter is neatly summed up in Guidelines for Creating Alternatives. Then, to consolidate your knowledge, there is a set of exercises and assignments which are carefully selected to help you improve your creative capacity.

TAPPING YOUR MEMORY

As you surely must know, you have in your own head a vast storehouse of knowledge and experience which is the equivalent of many, many books. Unfortunately, your access to this knowledge is relatively slow and erratic. Furthermore, if you search all of your knowledge and experience in a random fashion, it will take you an incredibly long time to find the answer. However, if you think over various aspects of the problem, it will *sensitize your connection to the relative knowledge in your brain.* For example, some people find that scanning trade journals will sensitize their memory to ideas they already know. It takes some pictures to sensitize the connections and bring the ideas into focus.

You don't need to trust your memory alone. You can *keep a record of design solutions* to difficult technical problems in your notebook. Have a means of indexing them for fast retrieval.

You can improve your retrieval of ideas by trying to *sensitize the route by which the information went into your noodle.* If it is a mechanical device, you can look at it, feel it, listen to it, smell it, taste it, regard it from all angles. These actions will often trigger off ideas buried deep in your memory. You can also make sketches, diagrams, and so forth and these actions will help you recall solutions to design problems which you already know.

WHO INVENTED THE WHEEL?

"Reinventing the wheel" is done so often that it is no longer a joke. It is simply a waste of time and talent. There is a vast storehouse of the world's experience you can tap.

Scientific progress is possible because knowledge can be shared with others by permanently recording it in print, drawings, photographs, computer files, etc. In order to develop an organized and systematic way of reaching the world's knowledge, use the four ways below.

FOUR WAYS TO TAP THE WORLD'S STOREHOUSE OF EXPERIENCE

1. Look Up the Graphically Recorded.

Look through textbooks, handbooks, journals, design records, catalogues and patent files. A little time taken in the beginning will save you much time and remorse in the long run. A busy engineer will have an assistant do preliminary scanning to bring ideas to his attention. Some organizations have found that the background research is so important that it is required before a new development or design proposal can get funding.

Find out what computerized abstract searching service is available to you. At the time of this writing, a search of five years of engineering journals costs only $30. This is one kind of service which will cost less every year.

2. Consult With Your Boss.

Your boss is obliged to help and usually has quite a storehouse of experience to call upon. Even when we know this, we frequently avoid asking for help for fear of appearing inadequate. I personally don't see why this should be an obstacle to getting ideas from your boss. When one of my subordinate engineers came to me for ideas, I was delighted to help. It made me feel like an important member of the team. The fact is that managers and subordinates need each other when it comes to producing results.

3. Get Ideas From Your Component Vendors.

They are very specialized on the applications of their components and they also are up to date on the best ways to use them. They are in touch with a lot more specialized technology than you are likely to be. I say this from experience as an applications engineer for semiconductors. By my contacts throughout the semiconductor industry, I knew of the latest techniques for better performance and lower cost. These I would give to any customer—if only they would ask. If they didn't ask, then I didn't consider it part of my job to interfere with their designs. So, pick away at that gold mine between your vendor's ears.

4. Benefit From Your Competitor's Efforts.

There's no use in reinventing the wheel. Look carefully at your competitor's products and find out all the good things that they have done. Then you can put your efforts into doing something better—where it counts. This advice is so logical and obvious that it hardly needs stating— yet in my experience in working with design engineers I have found many persons and many organizations who did not do this—professional pride, perhaps. But, remember, "pride goeth before a fall!"

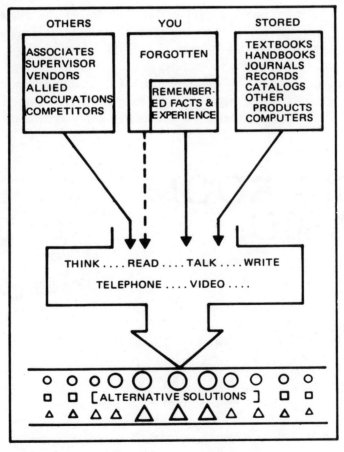

Figure 6.1. There are Many Ways to Tap the World's Storehouse of Experience.

IMPROVE YOUR GENERATION OF NEW IDEAS

Your own mind is the most creative tool you have. You need only open the gate and pour out a flood of ideas. Unfortunately for most of us, this gate is locked tight. As children, our parents and teachers derided us for our silly ideas. After fifteen to twenty years of formal education by means of regurgitating facts, the gate to new ideas is usually firmly closed. Figure 6.2 is a picture of a device that is now owned by tens of thousands of people. Yet, every time I show this picture on the screen to an audience, there is laughter. As adults, we still deride "silly" ideas.

There is a psychological explanation for closing the mind to internally generated ideas. If all the ideas that could be generated by combinations of experience were to pop into our conscious mind, there would be no chance to exercise our judgmental faculty.

In right-handed people the right side of the brain generates ideas and the left side of the brain provides judgment. The more you have in learned facts and experience, the better off you will be if you

REST ROOM RADIO

Solid state, full circuit toilet tissue dispenser radio . . .

for complete private enjoyment

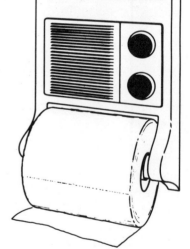

WHY DO WE RIDICULE NEW IDEAS?

RIDICULE   SCEPTICISM   SNEER   INSULTING

Figure 6.2.

develop the capacity to ideate with the right brain and apply judgment with the left.

We learn through education and experience to be very good at analysis. We become so good at analysis that we have a firm lock on the gate to our synthesis capability. Fortunately for us, however, a generation ago Alex Osborne showed us how to unlock this gate. He called it *deferred judgment.* Some people call it *brainstorming:* It is the simple *separation of analysis from synthesis.* When you wish to synthesize, that is, to get new ideas: withhold analysis, that is, defer judgment.

Most people who experience deferred judgment for the first time are surprised that they and their friends can produce so many ideas. Not all ideas are gems or even practical—but getting them all out is the way to come up with ideas. If you are looking for gold nuggets, you must look in the sand where you will also find tiny specks of gold, dirty lumps of mud, and rocks. That is to say, among a large list of ideas that you might produce by a private or group brainstorm, there will only be a few gold nuggets—nevertheless, that is the way to find the nuggets.

The so-called brainstorming technique can be used by yourself or in a group. Figures 6.3[1] and 6.4 are adapted from Alex F. Osborne's *Applied Imagination,* published by Charles Scribner's Sons, 1963. They are practical ways to use the brainstorming technique.

A word of caution: It's not enough to read about ideation, you must practice it. When you are proficient at getting ideas from your own mind, you may wish to develop the capacity to do this with groups. This requires a bit of special training. A lot of damage has been done to the reputation that belongs to brainstorming by attempts of untrained persons at leading groups in brainstorming. Find yourself an experienced leader in brainstorming and you will be glad that you did.

Whether you learn to brainstorm solo or in a group, there are always the social barriers to expressing new ideas. They are ridicule, scorn, and fear. Deferred judgment overcomes these barriers (See Figure 6.5).

[1] The SCAMPER idea is elaborated in the book: Eberle, Bob, *Scamper: Games for Imagination Development,* 1971 D.O.K. Publishers Inc., Radcliffe Road, Buffalo, New York, 14214

Substitute?	*Who else? What else? Other place? Other time?*
Combine?	*A blend? An assortment? Combine purposes? Combine ideas?*
Adapt? Alter?	*What is like this? What does it suggest?*
Modify? Magnify? Minify?	*Change material, color, motion, sound, odor, taste, form, shape? What to add? Stronger? Larger? Add ingredient? Multiply? Subtract? Smaller? Lighter? Split up? Less frequent?*
Put to other uses?	*New ways to use? Other uses if modified?*
Eliminate?	*Is it necessary? Is the use worth the cost?*
Reverse?	*Opposites? Turn it backward? Turn it upside down? Inside out?*

Figure 6.3. The SCAMPER Idea-Spurring Questions.

1. Criticism is Ruled Out:

Judgment is suspended until a later screening or evaluation session. Allowing yourself to be critical at the same time you are being creative is like trying to get hot and cold water from one faucet at the same time. Ideas aren't hot enough—criticism isn't cold enough. Results are tepid. (Do not analyze while you synthesize.)

2. Free-Wheeling is Welcomed:

The wilder the ideas, the better. Even offbeat, impractical suggestions may "trigger" practical suggestions which might not otherwise occur.

3. Quantity is Wanted:

The greater the number of ideas, the greater the likelihood of winners. It is easier to pare down a long list of ideas than puff up a short list.

4. Combinations and Improvements are Sought:

In addition to contributing ideas of their own, panel members should suggest how the ideas of others can be turned into better ideas, or how two or more ideas could be combined into an even better idea.

Figure 6.4. Osborne Rules for Brainstorming Sessions.

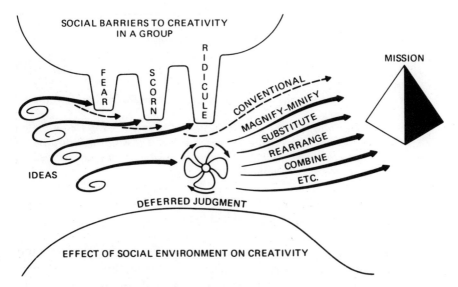

Figure 6.5. Effect of Social Environment on Creativity.

CREATE BY "SLEEPING ON IT"

The process of *illumination* is a familiar one. An idea suddenly pops into your mind like the proverbial lightbulb. You may be working and thinking about another problem, or you may be sitting in a movie, or you may be sleeping while the idea is *incubating.* The ideas can be good. You must be prepared to note them down—or else, like moonbeam reflected in a cat's eye, they may be gone forever.

Psychologists have studied this illumination and have developed a technique. The rules are in Figure 6.6.

ANOTHER WAY TO EXPAND THE DESIGN ALTERNATIVES

Morphological analysis is a way to determine the substructure of a design problem and to force yourself to think of oddball combinations. Some engineered designs can be broken into functional subsystems as demonstrated below in Figure 6.7. There we see the creative alternatives for three of the subsystems for an electric toothbrush.

1. *Work hard on the problem until you cannot come up with any other solutions that satisfy you.*
2. *Put the problem completely out of your mind and do something else.*
3. *Keep a note pad handy with you wherever you are, including while you are sleeping.*
4. *If the bright idea doesn't appear within 24 hours, work again on the problem and let your mind incubate it again. The result will surprise you. What comes naturally only once in a while, can be developed into a made-to-order technique.*

Figure 6.6. Incubation and Illumination Rules.

To proceed with the morphological analysis, having constructed the morphology as in Figure 6.7, take each item in the first column and combine it with every item in the second column, and again with every item in the third. Many of these are impractical combinations, but since we are looking for useful ideas, it is not necessary to make all combinations. What we wish to do is to trigger your mind into new ways of thinking. For example, you can combine "brush

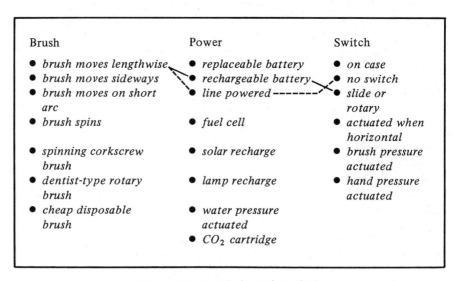

Figure 6.7. Morphological Analysis.

moves lengthwise" with "rechargeable battery" and with "rotary switch." Another possibility is "brush moving lengthwise" with "powered from line" and with "no switch." The morphological analysis does not come up with anything fundamentally new, but it does direct your mind to unusual combinations of otherwise ordinary things. Your payoff comes from discovering better alternatives for the situation at hand.

INVENT BY THE PERSONAL ANALOGY TECHNIQUE

Deferred judgment, incubation and illumination, and morphological analysis are tried and proven ways of generating ideas for minor design improvements. They have not, however, proven of great value for invention. Creating something absolutely new for the whole world is something that doesn't happen very often—even with inventors—and the process is not well understood. Good design is only part invention—most design success comes about through making performance improvements on some basic idea. Nevertheless, there are two techniques I would like to point out for cases where you wish to stimulate invention. They are the Personal Analogy Technique and the Idea Trigger Session.

The first of these is the Personal Analogy Technique explained by J. J. Gordon in his book on *Synectics.* You stretch your imagination to the utmost and imagine that you are part of the thing that is being designed. You are then able to perceive qualities and attributes of the thing you are designing that you would be blind to otherwise. Remember the old Indian proverb, "You do not know another man until you have walked in his moccasins." You do not understand a design until you have been a real part of it. Of course, this can only take place in fantasy.

An example will clarify the technique. One person plays the role of the car and the other person plays the role of the gasoline pump. "Hi, Joe. How're you feeling today?", says the pump. The car replies, "Well, not so good. I don't know whether I want high-test or regular gasoline. Just mix me up a cocktail that will pick me up." Thus was born the idea of blended gasoline.

A personal analogy that I demonstrate in my seminars is the analogy of being a tire which is forced around a corner. I hold on to a chair and with my feet I grip the floor like a tire and try to make a sharp turn. I'm thinking and saying, "My feet want to slip and I can't make the turn. If only I could keep my feet flat on the floor with rubber soles, then I could make the turn easier. Thus was born the idea of the radial tire which flexes and keeps the rubber surface on the road.

This demonstration usually gets some laughs because such behavior among engineers appears a little bit ridiculous. So devastating is the power of ridicule that most people will not try this method unless encouraged by a group of workmates who also want to learn the technique. Then they practice and do it in private. They don't have to tell the "straight" world where they get their ideas.

THE IDEA TRIGGER SESSION *good for generating potential ideas*

In this technique, everyone is sober and serious. They are put in a situation where they think very hard to get new ideas to solve a problem. The Idea Trigger Session is helped by the use of a form and process developed by George H. Muller. This is shown in Figure 6.8. Note the direction at the top of the form "use key words only." This is important and that is the reason for the narrow columns in Muller's form.

The idea of a private and silent brainstorm with mental stimulation from your own output is a technique I learned at a seminar in 1962. I'm inclined to give credit to William J. Osborne for the private brainstorm technique. However, George Muller has taken it a step further. One's private and silent brainstorm is stimulated by the spoken and visually recorded *new* ideas of other members of a group of experts. The new ideas are taken up in an orderly fashion. The participant is not bothered with a mountain of trivia. There is time to think. The name of the game is to get an idea so new that no one else has presented it. It works. See Figure 6.9 for the steps in the Idea Trigger Session process.

Inventions are attributed to the process of the Idea Trigger Session. One of these is an energy absorbing hub/shaft system used on some

THE IDEA TRIGGER SESSION

CONTRIBUTOR _____

USE KEY WORDS ONLY.
MOVE SWIFTLY FROM
ONE IDEA TO THE NEXT

IDEA TRIGGER SESSION NO. _____
3 EXCURSIONS

FIRST	SECOND	THIRD	NOTES
120 SEC.			
30 SEC.			

TOTALS

ORIGINALS _____   _____   _____   _____

DUPLICATE _____   _____   _____   _____

TOTALS   _____   _____   _____   _____

Figure 6.8. Copyright © 1968 G. H. Muller.

THE IDEA TRIGGER SESSION[1]

Silent Phase or Purge—*In the silent idea production phase, the obvious ideas are produced by the participants who write* quickly *in the first column of Figure 6.8. Since tension, or relaxation, or alternatively applied pressure and relief, have been found to enhance creativity, two minutes are given by the* operator *to write idea key words in the first column, while participants are reminded by the operator's voice, "One minute to go," "Fifteen seconds to go" of the time left. After a few minutes relaxation, they are given another 30 seconds to complete the first column. "Blue sky" ideas are not discouraged since they may trigger useful ones, later in the Spoken Phase. Ideas must be expressed keeping in mind that the subject at hand (also called want or market vacuum) is the result of a disequilibrium between the actual customer behavior pattern, his expectations and the current technological answer. At the end of the* Silent *Phase participants have purged themselves of all the ideas that came to mind.*

Spoken Phase or Trigger—*In this orderly phase, and for each trigger excursion, each participant, one at a time, clockwise, describes rapidly his ideas. The others* silently *cross off their lists duplicate ideas they may have, and add new ones continuously in the second column as they are triggered by the ideas of the speakers. If participants are slow in filling the second column, the operator may announce that he will at times and on a random basis request from the speakers, as they describe their first column ideas, to also describe the third idea that they may have in the second column as a result of being triggered. Thus the schedule of the process automatically assures that trigger takes place. After the first excursion, a second one follows counterclockwise, using second column ideas, while new ideas are written in the third one. Second and third column ideas are the most important since they probably were not present at the onset in the deep recesses of the participants memory tracks, or could not have been formulated by associations or analogies due to lack of foreign input. The ideas of the second and third columns are mostly the result of the trigger phenomenon because:*

- *Either one is miffed, that is,* negatively *reinforced for being duplicated and moves in a new direction.*
- *Or one is* negatively *reinforced by competitive pressure, willing to*

[1]This description is an adaptation from appendix I of THE IDEA TRIGGER SESSION PRIMER, copyrighted 1973 by George H. Muller, available from the A.I.R. Foundation, 2921 Overridge Dr., Ann Arbor, Michigan 48104.

> *do better (upmanship) or differently (originality) than others just heard.*
> - *Or one is positively reinforced by a substantial list of one's own not yet duplicated ideas and is adding more.*
>
> *A student group will take one hour for the first time per excursion, 30 minutes after training. Few ideas are left for a fourth column and the group is generally mentally exhausted at this point. Throughout the process, the operator, among many duties, asks questions to the speaking participants to assure by their answers that ideas are understood and obvious duplicates eliminated.*
>
> *On difficult technical problems, the purge phase averages 8 minutes, and the trigger phase may last 4 hours.*

Figure 6.9. The Idea Trigger Session.

Ford car models, described in "TECHNA—A concept car to challenge automotive engineers," G. H. Muller, *Society of Automotive Engineers Congress*, Detroit, Jan. 13–17, 1969, paper 690267. A brief description of the Idea Trigger Session was carred in *Management Review*, Sept. 1977, pp 35 and 36.

HOW MANY ALTERNATIVES SHOULD YOU HAVE IN A DESIGN?

For a design which is not particularly innovative, such as a common truss with a new set of dimensions, a commonplace electronic circuit, a conventional gearbox to meet special requirements, and so forth, you know what the answer is. The objective is to get out a reliable design as soon and as cheaply as you can. For this situation, one alternative or possibly two or three slight variations are all that you need in order to know that you are on the right track.

But what about innovative design? When do you know that you have "the best" alternative? Never! The best solution may not be known for a hundred years—but you can get quite close with an optimal-type solution selected from a large range of alternatives.

The quality of the alternative that you go ahead with is judged by its comparison with other alternatives. With innovative design, it is important that you go with the alternative which is quantitatively and subjectively better than the conventional alternatives. In such a situation, it is not unusual to make a shopping list of ten to twenty alternatives, including some fairly wild ones. This list is pared down by selection of those which are practical under the circumstances. Frequently five alternatives may be analyzed for cost and technical feasibility, and three may be developed out to the final decision point.

Generating ideas is a divergent activity. You are on the right track when you get more ideas than you can possibly use. This is followed by a convergent activity which is to check out the feasibility of the ideas and to make a selection on some logical basis. This convergent activity will be the subject of the following two chapters.

THE PRINCIPLE OF CREATING ALTERNATIVES

All of the techniques described above can be generalized into one important thought. You defer judgment while you are ideating. You do not analyze while you synthesize. The judgmental side of the brain is held back while the imaginative side freewheels. In a generalized form, we can state it this way:

THE PRINCIPLE OF CREATING ALTERNATIVES.

When you want to get new ideas for design alternatives, use a technique that defers judgment until ideas are generated.

APPLICATION TO AN EQUIPMENT DESIGN

The design of an electric power transformer is straightforward handbook design in most cases. To look at the creative side, consider a basic redesign for a special application.

A Substation Transformer: Alternative Design Concepts

Existing Alternatives

Windings:

- *Copper*
- *Aluminum*
- *Solid, stranded, foil, or transposed*

Core:

- *High or low loss*
- *High or low permeability*
- *Thick or thin laminations*

Insulation:

- *Oil and paper*

Creative Alternatives

Windings:

- *Cast copper or aluminum*
- *Hollow conductors*

Core:

- *Rebuilt core*
- *Tape wound core*
- *Ceramic core*
- *New high density, material x*

Insulation:

- *Ceramic*
- *Silicone*
- *Gas*
- *Solid impregnant*

By using a morphological analysis, the above components can be combined in many ways and examined for ideas of practical value.

APPLICATION TO A CONSUMER PRODUCT

If you really want new ideas for an innovative electric toothbrush, you may want to use the technique of brainstorming followed by a morphological analysis. You will determine the subsystems and the alternatives for each. You will combine these in the way shown in the earlier text. This will give you a wide range of possible innovative designs.

The result of my own private brainstorm is shown below.

An Electric Toothbrush: Alternative Design Concepts.

Existing Alternatives

  Brush:

- *Brush moves lengthwise*

- *Brush moves sideways*

- *Brush moves on short arc*

  Case:

- *Metal*
- *ABS plastic*
- *PVC plastic*
- *Metal and plastic*

Power:

- *Powered by replaceable battery*
- *Powered by rechargeable battery*
- *Powered from line*

Switch:

- *On case*
- *No switch*
- *Slide or rotary*

Creative Alternatives

  Brush:

- *Brush spins*
- *Spinning corkscrew brush*
- *Dentist type rotary brush*
- *Dentist type rotary rubber*
- *Replaceable fabric*
- *Cheap disposable brush*

  Case:

- *Fiberglass reinforced plastic (FRP)*
- *Wood*
- *Wood veneer*
- *FRP and metal*

  Features:

- *Toothpaste in handle*
- *Permanent abrasive in brush*
- *Cocktail mixing head option*
- *Adjustable speed*

Power:

- *Fuel cell*
- *Atomic recharge*
- *Solar recharge*
- *Lamp recharge*
- *Water pressure*
- *CO_2 cartridge*

Switch:

- *Actuated when horizontal*
- *Brush pressure actuated*
- *Pull out line cord*
- *Hand pressure*

APPLICATION TO A LARGE-SCALE
CAPITAL PROJECT

Ordinarily, a hydroelectric dam would not require much creativity, standard practices would probably do. However, in this situation we have the additional requirements of an irrigation system and a symbol of national pride. In the beginning, it would be a good idea to explore all possible alternatives for the hydroelectric dam and irrigation system. Morphological analysis is appropriate here because of the way it breaks down into subsystems. The only really difficult problem requiring a unique solution is obtaining a symbol of national pride at very little cost. Deferred judgment, such as the brainstorm or incubation and illumination, would be appropriate here. In the example shown below, I have gone from the usual alternatives which would be suitable for morphological analysis, to a list of creative alternatives which would be suitable for a conceptual design of something really different.

A Hydroelectric Dam for Latin America: Alternative Design Concepts.

Usual Alternatives

Dam:

- *Earth*
- *Stone*
- *Concrete*

Power:

- *At top of dam*
- *Below dam*
- *Distant from dam*
- *Large turbines*
- *Small turbines*

Irrigation:

- *Gravity*
- *Pumped*

Standby Street Lighting:

- *Diesel*
- *Gasoline*
- *Turbine*
- *Line to other source*

Creative Alternatives

Dam:

- *Prestressed concrete*
- *Compressed timber with resin*

Power:

- *Under dam*
- *Under water*
- *Side channel*

- *Compressed garbage*
- *Masonry*
- *Keyed blocks*

- *Decentralized*
- *Downstream*

Irrigation:

- *Canal from upstream storage*
- *Solar powered pumps*
- *Wind powered pumps*
- *Subsurface root wetting*

Standby Street Lighting:

- *Storage batteries*
- *Kerosene lanterns*
- *Propane lanterns*

Symbol of National Pride:

- *Widest, deepest, or highest*
- *Unique materials*
- *Dam embossed with national bird*
- *Give prize for local sculpture*
- *Build shrine pieces into dam*

A CASE HISTORY OF AN EQUIPMENT DESIGN:
A DRIVEWAY SNOWPLOW

A group of mechanically oriented designers were trying to design a device to clear a home driveway with the utilization of an automobile. Their first set of alternative solutions was as follows:

A. Fixed. Device is fixed to the car and the car moves:

1. plows
2. augers
3. blowers
4. heating elements
5. scooping

B. Portable. The car remains stationary and the device operates by power from the car.

1. electrical
2. pneumatic
3. hydraulic
4. mechanical

Later, another meeting was held at which they *brainstormed* the problem and came up with the following alternative solutions (in orderof naming):

1. Push-a-plow
2. Pull-a-plow
3. Two-stage plowing
4. Drive auger
5. Combine an auger with a rotary brush
6. Single-angle plow
7. Double-angle plow
8. Front-end loader
9. Rotary overlapping scooper
10. Blower
11. Heating elements
12. Combination blower and heating elements
13. Roll the snow flat
14. Conveyor belt
15. Dredger bucket
16. Use exhaust gases for melting
17. Use rear wheels on rollers for drive
18. Use No. 17 to push a plow
19. Push an auger
20. Vacuum-drive
21. Hot water spray
22. Combine an auger and a blower
23. Use a pulley system to pull a skid, using No. 17 for drive

Comments: Look at the list of 23 ideas resulting from their brainstorm. You will probably agree that plows, augers and brushes are fairly conventional solutions to removing snow. On the other hand, using exhaust gases, a conveyor belt, hot water spray, or using the rear wheels on rollers for drive are novel solutions. So one might say that the real novelty began at 11 or 16. The first part of such a list is usually a purge of conventional ideas, and some people have observed that it takes about 17 ideas to reach the really novel ones.

Did the designers practice Osborn's rule of freewheeling? The wilder the ideas, the better, according to Osborn. Well, using the exhaust gases for melting, or a hot water spray are a bit wild, wouldn't you think?

Did quantity breed quality? Of course, this question is subject to a high degree of personal judgment. The final design used by the group was a plow fixed to the front with additional traction obtained by the rear wheels moving a roller-driven traction device. Apparently, the 17th idea led to the final design solution.

Other evidence of quality coming from quantity is that the second list is a considerable expansion of the original list which had only five solutions. It took more than three times this number to arrive at the novel solution which was considered satisfactory.

DESIGN GUIDELINES

GUIDELINES FOR CREATING ALTERNATIVES.

Do	Don't
Do tap the world's storehouse of experience for alternatives.	*Don't ignore the experience of others even if you may someday prove that you are a genius.*
Do try to get enough alternatives so that you can better judge the one to go with.	*Don't run off with the first and only idea that comes to your quick mind.*
Do turn your judgment off when you try to create alternatives.	*Don't freeze your mind with criticisms of your own ideas.*
Do accept other people's ideas when you want more ideas to come forth.	*Don't crack up when your colleague comes up with a wild idea.*
Do use some ideation technique when you seek improvements or innovations.	*Don't imagine you are creating when in reality your ideas are in deep freeze.*

HOW TO BECOME MORE CREATIVE

Will reading this chapter make you a more creative person? Not likely. If reading a book could develop creative capacity, then books alone would do it. Creativity is a behavior that you were born with, but which became suppressed to a greater or lesser degree by your life's experience. To be more creative, you must learn to generate and release new ideas. You must practice. As a matter of fact, even creative persons must practice, or society will gradually turn them off. Don't depend on serendipity. If you don't start now to practice creative techniques, you will be about where you were before.

You will see in the questions and assignments that follow, that there are things you should do by yourself and things you should try with a group. All of these are designed to lead you to be a more creative person.

EXERCISES AND ASSIGNMENTS TO HELP YOU
BECOME MORE CREATIVE

Questions on the Contents of this Chapter

1. Imagine that you must put the quintessence of this chapter into a cablegram of ten words or less. What would you say?
2. Which of the techniques described are good for modest design improvements?
3. What are the two techniques described which have a track record of encouraging invention?
4. Refer to the "Application to a Consumer Product—An Electric Toothbrush: Alternative Design Concepts." If morphological analysis is used to combine all possible subsystems, what is the approximate number of different combinations which could be examined?
5. For getting alternatives, this chapter lists four ways to tap the world's storehouse of experience and five mind-expanding ways to create new ideas. Which are most appropriate for:

a. the design of prefabricated trusses which are to be mass-produced from lumber, pre-cut and set in jigs?

b. the design of a prefabricated modular structural element from a light-weight alloy for use in helicopter transported prefabricated dwellings?

c. the development and design of sensors for a satellite which will test for life on Mars?

d. the redesign of a farm tractor for the purpose of reducing its manufacturing cost to the minimum required for a utility tractor which will perform as well as the present model?

Questions on the Case History

1. Examine the list of 23 brainstorm ideas. By listing number sequences, show how one idea led to another in at least six instances.

2. How many really wild (and impractical) ideas are in the list?

Assignments to Help You Become
a Super Creative Pro

1. For your practice project, or your real project, there will be some alternatives to look at. List some possibilities such as: alternative basic concepts, alternatives for some of the subsystems or components, ideas to reduce the cost of one component, ideas to overcome a technical problem and so forth. There are five mind-expanding ways in this chapter and four of these you can do by yourself. Work your way through these methods and indicate the appropriateness for listing possibilities, ideas, alternatives, etc. for your project. Then give some time to practice with those techniques which are new to you. If you are working with a group on the project, you can try the techniques of group brainstorming and the Trigger Idea Session.

2. During your next working week, try to improve your tolerance to new ideas and this will improve your personal creativity. For example, try driving to work on a different route for each day

of the week to see what you can discover about yourself. Try ordering strange, exotic or even repulsive foods from a restaurant menu. Think of them as new ideas which require tolerance on your part.

3. Try to work some creative activities in your social life, such as having weekly brainstorm sessions just for the fun of it, creating jazz music, creative dancing, or story telling.

4. Walk around your work environment and imagine creative improvements that you would make in the facilities if you were permitted to do so. In your mind's eye, rearrange the machines, the buildings, the furniture and small items so as to practice your creative behavior. Being creative and remaining creative is an uphill struggle and you should practice continually.

7 How to Check Out
the Feasibility of a Design Idea

KEY IDEA: A DESIGN IDEA IS FEASIBLE ONLY WHEN THE PER-
FORMANCE, COST, AND TIME OBJECTIVES ARE PRACTICAL
UNDER THE CIRCUMSTANCES THAT PREVAIL.

IT'S A GOOD IDEA, BUT . . .

A few years back, before our major energy crisis, a team of engineering and anthropology professors from the University of Chicago
tried to improve the material way of life in a Mexican village. The
problem to be solved was how to generate an adequate supply of
very hot water for dyeing the wool used in their blanket-making
industry, without denuding all the nearby hills of their forests for
fuel. The engineers came up with the bright idea of using solar heat.
They designed an easily made unit with a parabolic reflecting surface
which could be manually adjusted to point at the sun. Technically,
it worked fine. There was plenty of sunshine. It was technically
feasible. Was it economically feasible? Perhaps. It was so designed
that local craftsmen could fit it together for a sum that a local family
might afford. In the meantime, since they were given away to Mexican families for their experimental use, they were economically and
financially feasible. What about the timing? It would be available in
plenty of time because wood was still available within ten miles of
the village and was not expected to run out for a few more years.

132

So far so good with the solar heaters. You might expect that these would have been a resounding success, and so they were insofar as the experiment went. However, the research showed that the equipment needed to be socially and culturally feasible as well. What happened was this: In this village, the women were responsible for the cooking and therefore for the boiling of the water. They did not, however, share the technological inclination of the male head of the house who elected to use the machine. They did not understand why the unit had to be polished clean everyday and they did not understand the idea of tracking the sun. Moreover, they still had to have a wood-burning heater for days when it clouded over. Two years later, only one of the units remained in operation. What the researchers concluded was that technically they had a good idea but that **under the circumstances it simply wasn't a feasible design solution.**

HOW THIS CHAPTER WILL ASSIST YOU IN YOUR DESIGN EFFORTS

Objective

You should learn how to take a long list of design alternatives and reduce it to a short list of alternatives which are worth considering further.

Benefit

You will save yourself design time by pruning 'idea' branches which look interesting but will not get you anywhere under the circumstances.

This fifth step, *Screen for Feasibility* follows the step *Create Alternatives* of the preceding chapter and comes before the step *Select the Solution.*

In this chapter, you will be given specific checklists to guide you on assessing Technical Feasibility and Resource Feasibility. You will also learn how to check out the basic practicability of a design idea for its timing, human factors, social, and cultural factors. You will be shown how to estimate the costs and financial return on a consumer product.

Then, in the applications, you will see how to construct a *Check-out Feasibility Table.* The examples are useful models for you to use for engineered designs where feasibility is more than just a technical question.

Guidelines to Screen for Feasibility sums it all up compactly for your use on real design projects.

The follow-up exercises will assist you to improve your understanding of the step, *Screen for Feasibility.*

FIRST, APPLY CRUCIAL CRITERIA
AND RESTRAINTS

Some design alternatives are clearly not feasible because they do not meet some physical or legal restraint. I like to refer to these as crucial criteria because although they are criteria for acceptance, they are go or no-go in their application.

An example of a go/no-go crucial criteria is that a washer-dryer must be able to pass through the doorway of a house. This is a very practical consideration.

The reason for applying the crucial criteria in the beginning of this step is that you can eliminate many alternatives from further detailed analysis and thereby save yourself some effort. Once this is done, the remaining alternatives can be looked at in a framework of examining the performance, cost and time in the circumstances that prevail.

WOULD IT WORK UNDER THE CIRCUMSTANCES?

A technical feasibility analysis is to determine whether or not the performance level would be sufficiently close to the original objectives. Knowing the user needs, and the tradeoff they are willing to make can help you make a decision about technical feasibility. Often a wide range of technical outputs will be satisfactory. Consider the design of automobiles, bicycles, household appliances, conveyor systems and bridges. We engineers are seldom in any real difficulty here unless it is a highly innovative design. We exercise a knowledge of our discipline and only propose those things that will work—well, usually.

A CHECKLIST FOR TECHNICAL FEASIBILITY.

1. *Are any of the laws of physics being violated?**
2. *Will the expected performance level be adequate under the circumstances?*
3. *Can it be made with the facilities, talent, and machinery available to the design sponsor?*
4. *Will it operate reliably under the conditions of use or abuse?*
5. *Will it meet the minimum required in the technical specifications?*
6. *Will there be interactions between subsystems that could give trouble?*
7. *Can the machine or equipment be used with comfort and convenience by the people who must interact with it physically or mentally?*
8. *Can it be serviced and maintained with reasonable cost and dispatch? Will parts be available?*
9. *Is the performance level good enough to compete with competitive offerings?*
10. *Is the technology within the scope of the people who must build and maintain it?*
11. *Are there any spurious outputs of concern?*
12. *Are there any unusual extremes of temperature or moisture in its use?*
13. *Is it really safe to use? Is it foolproof enough, or does it need to be damfoolproofed?*
14. *Are there any airborne noises or chemical pollutants of concern?*
15. *Are the esthetics important, and if so, what comparisons are there?*
16. *Is it adaptable to future changes?*

---

*For a table of 113 laws of physics, see pp. 75–80 in T. T. Woodson. *Introduction To Engineering Design*. New York: McGraw-Hill, 1966.

ARE THE RESOURCES AVAILABLE?

Some designs take more resources in money and talent than others, and may exceed that which is readily available. No organization has infinite funds for financing projects, and a lack of available expertise is a consideration when money can't buy it.

A CHECKLIST FOR THE RESOURCES FEASIBILITY.

1. *Is there enough capital available to finance the whole project from design start until the product or equipment is sold and the investment recovered? In the case of the government or a corporation with large capital resources, this question must still be answered because of competing proposals for the use of the available capital (the financial feasibility).*
2. *Is the cost of making a solution in proportion to the benefits to be gained? For a consumer product this is simply a profit calculation. In the case of a large-scale socio-economic system like a hydro-electric dam, there may well be other benefits than the technical ones. The so-called "economic viability" of such a project must be done with cost-benefit or cost-effectiveness analysis (the economic feasibility).*
3. *Is the expertise available for the amount and timing that could be expected?*
4. *If appropriate, would it be cheaper to buy it than to make it?*
5. *If appropriate, is the return on investment and profit per unit up to the standards required?*

Let us look at an example of calculation for a ***financial and economic feasibility analysis.***

This is a calculation based upon the design of a portable TV with which I was once associated. The pattern shown here will be useful to you for any similar kind of calculation on a product or equipment for which a profit must be shown.

Design and development cost	$50,000
Tooling cost for plastic cabinet and parts	$25,000
Modifications to manufacturing facility	$ 5,000
Total investment	$80,000
Inventory cost for materials and labor to make first production run of 1000 units at $100 each	$100,000
Total cash needed before sales income	$180,000

From the above, you can see that the entrepreneur must have

$180,000 available through borrowing or reserves in order to "put the show on the road." If he has it, then the project is financially feasible, if he hasn't, he may go bankrupt. That is the fate of many small entrepreneurs who do not make this calculation.

What do the shareholders get for their investment of $80,000? Suppose that they normally get 15 percent after taxes and that this corresponds to 30 percent before taxes. To simplify matters, let us suppose further that the $80,000 is made available at the beginning of an eight-month design period and recovered on 1000 units.

The desired return on investment (ROI) is equal to:

$$\frac{30\% \times 8 \text{ months}}{12 \text{ months}} \times \$80,000 = \$16,000$$

This corresponds to $16 per unit to be sold. We will now make a calculation of the selling price for such a product.

Factory cost for material and labor, same as inventory cost above	$100.00
Plus recovery of investment of $80,000 on 1000 units*	$ 80.00
Plus return on investment expected (profit)*	$ 16.00
Equals cost to sales department	$196.00
Plus sales markup to cover selling, advertising, and warranty, 20%	$ 37.20
Equals selling price to distributor	$233.20

(Profit as a percent on the corporation's selling price is

$$\frac{16}{233.20} \times 100\%$$

which is approximately equal to 7%)

Plus distributor markup, 15%	$ 33.60
Equals distributor selling price	$266.80
Plus dealer markup, 40%	$102.70
Equals final retail selling price	$369.50

*An accountant would use amortization tables.

From this calculation, we can see that the new design is economically feasible if the users will pay about $370. It is financially feasible if the organization can manage to have $180,000 in cash for a period of about eight months. If more than 1000 units were made, the recovery of the investment could be less per unit and the selling price could be reduced accordingly.

CAN IT BE COMPLETED ON TIME?

Timing Feasibility

The best time for completion of a design project is usually determined by outside circumstances. It is not, as is commonly expressed, the time taken to get it completed. In the case of a product there is an opportune time for being on the market, considering such things as seasonal variations, competitive efforts and maximum long-term investment. It is all very well to say that it was needed yesterday. It is sometimes possible to purchase someone else's design and get it out in a hurry, but it may be wiser to take longer and get a technological edge over the competition.

What is the right timing for a public works project such as a hydroelectric dam or a highway? If we rush it through as a crash program, the capital cost will be undoubtedly higher than if we can take time to explore design alternatives and use contractors when they are hungry. But if we take longer there is some social and economic cost for it not being in place. In many instances this can be offset by the long-term benefits of taking the time to do a very good job.

Not all design solutions can be completed in the same time. Some of the more esoteric ones generally take longer. For example, a design of a TV set with a competitive edge will take longer than an update of last year's model. The design of a huge machine to cut down and harvest trees in the forest will take longer if it is better than anything already on the market. These are the things we must consider when deciding among alternatives with respect to the time taken. Note, however, that the main considerations are the external forces which determine when it is needed, rather than the design-generated forces that determine how long it will take to complete it. When these times are in conflict, a design program may be rushed through on a crash basis, provided that additional human and financial resources are made available.

WILL IT BE ACCEPTABLE FOR THE BODIES AND MINDS OF THE USERS?

Human Factor Feasibility

The users must interface with machines and equipment in some fashion. In many cases, such as an automobile, there is a strong physical connection that cannot be ignored. The study of this is called human factors engineering or ergonomics. There are books with tables which show the limitations and strength of all the human limbs, all the common dimensions, and the statistical range. These books contain all kinds of suggestions for the visual layout of panels. Human factors engineering has become a specialty discipline with its own society.

At this point in the design process, we are really only interested in the feasibility of a number of alternatives. When we choose one we will make sure that the human factors are looked after, so that when screening for feasibility it is only necessary to look for any unusual human factors that might be generated by one of the new ideas. If you venture into unknown territory, consult a human factors specialist because there are some things that you don't generally come across. For example, did you know that a visually alternating pattern around three cycles per second can trigger an epileptic fit in a susceptible person? Sound levels above 120 decibels can cause permanent ear damage, pinpoint light sources can damage the eye retina, and moving objects within an operator's peripheral vision can cause nervous tension. So talk to the specialists. It is important.

Social-Cultural Feasibility

The social-cultural feasibility pertains to what people think about the design solution. Some of the thinking is already embodied in laws regarding environmental pollution and safety standards. However, in this day of rapid social change, one must be alert for public acceptance of things which may as yet not be covered by laws. This is why industrial buildings must often have a pleasing landscape. This is why the most efficient overhead conveyor system may not be acceptable for noise and appearance when near a residential area. People have concerns about esthetics not only relating to buildings, but also relating to the machinery with which they work. Buyers of industrial equipment are influenced by esthetic factors even when profit is supposedly their topmost concern.

A PRINCIPLE OF DESIGN

THE PRINCIPLE OF FEASIBILITY.

A design alternative is feasible when the expected levels of performance, cost, and time are practicable under the special circumstances in which the design will be developed and used.

APPLICATION TO A CONSUMER PRODUCT

An Electric Toothbrush: Some Factors to be Considered When Screening for Feasibility with the Alternatives Suggested in Chapter Seven.

Crucial Criteria

UL standards would not be met if a radioactive material was used.
An imposed limit of $50,000 for tooling costs could be met by all the alternatives.

Technical Feasibility

Atomic recharge is not technically feasible for a low voltage battery (but this idea led to others).
All others are within the technological capability of most design organizations.

Resource Feasibility

Economics Factors—Fuel cells, solar recharge and lamp recharge would cost more than people would pay ($12.50 to $17.50).

Capital and Expertise Factors—All are within the resources of a large corporation, but a small company would not have the resources to develop and test some of the creative alternatives. We will proceed as if a "small" company is the entrepreneur, and this is the reason that the maximum tooling cost cannot exceed $50,000.

Human Factor Feasibility

Children will probably have difficulty with a switch which actuates when horizontal. A human factor study is needed to find out a child's hand pressure if a pressure actuated switch is used.

Social and Cultural Feasibility

There would be a cultural barrier to overcome if anything much different from usual were used. Next in acceptability would be something like a dental rotary cleaner, since dentists are highly respected in our society. This is all the risk our "small" company could be expected to take.

A solid color plastic case would be more culturally acceptable than wood or polished metal, due to the North American trend to solid color plastics or ceramics in bathroom decor.

You can put all the feasibility tests into a neat arrangement called the *Quick Check-out Feasibility Table.*

A few of the possible alternatives for the electric toothbrush are used in the table shown in Figure 7.1.

APPLICATION TO A CAPITAL PROJECT

A Hydroelectric Dam for Latin America

On such a project, the technical feasibility is seldom the question. What counts is the impact it has on the society or organization that it is supposed to serve. In this example, I have chosen two alternatives which are both technically feasible, but have other problems

ALTERNATIVE → ↓ FEASIBILITY	A LINE-RECHARGEABLE-BATTERY; BRUSH MOVES SIDEWAYS	B LINE POWERED; BRUSH MOVES SIDEWAYS	C REPLACEABLE BATTERY; DENTIST-TYPE ROTARY RUBBER TIP	D SOLAR/LAMP RECHARGE; BRUSH MOVES SIDEWAYS
CRUCIAL CRITERIA Saleable in U.S.A. & Canada by meeting UL & CSA Safety Standards	OK	OK	OK	OK
TECHNICAL FEASIBILITY (Performance) Attractive; functions as good as Brand X	OK	OK	OK if brush used for rear surfaces	OK
RESOURCE FEASIBILITY (Cost) Selling price not more than 10% more than present utility models	OK	OK	OK	Too expensive and energy saved does not offset the cost
TIMING FEASIBILITY (Time) Ready for Christmas trade	Would miss first season because of development time	Would miss first season because of development time	OK	Would miss first season because of development time
HUMAN FACTOR FEASIBILITY Child can use	OK	OK	Requires training of children	OK
SOCIAL-CULTURAL FEASIBILITY Hygienic; personal use or gift item	OK	Some concern over shock hazard	Would appeal to those who want to polish their front teeth	Sales would benefit from user being energy conscious

Figure 7.1. The Screening for Feasibility Alternatives for an Electric Tooth-brush with a *Check-out Feasibility Table*.

to be overcome if either is to be practicable. See the *Check-out Feasibility Table* of Figure 7.2.

Comments: Think of the number of capital projects in trouble today because of the social-cultural factors. We can't find sites for electric power plants, for pipelines, for harbors, etc. Environmental and social concerns are important. Be sure to take a good look at them on capital projects.

APPLICATION TO AN EQUIPMENT DESIGN

The substation transformer used as an example previously is not worth the effort of completing a feasibility table, because most alternatives would be viable.

ALTERNATIVE → ↓ FEASIBILITY	SITE A A HIGH DAM	SITE B A LOW, WIDE DAM
CRUCIAL CRITERIA Foreign aid contract restriction	OK	OK
TECHNICAL FEASIBILITY Adequate electric power and irrigation water	OK	OK
RESOURCE FEASIBILITY Installed for X millions	OK	OK Substantially lower in cost than A
TIMING FEASIBILITY Completed by Jan. - -, 19 - -	OK May be late, but tolerable	OK
HUMAN FACTOR FEASIBILITY Man–machine interface	OK	OK
SOCIAL AND CULTURAL FEASIBILITY Preservation of National Shrine	OK Costly to protect shrine, but it can be done	NOT OK Would flood villages and farms of militant minority of 10,000 persons

Figure 7.2. The Screening for Feasibility Alternatives for a Hydroelectric Dam with a *Check-out Feasibility Table.*

What to watch out for on an equipment design is the Human Factor Feasibility. The people who use the equipment must be able to operate the controls and service it. Hands-on-equipment should be convenient to use in order to be fully viable in its context of use.

APPLICATIONS IN GENERAL

From the examples, one can see that the technical feasibility of consumer products is seldom a matter for concern. Nevertheless, a manufacturer must have the technological capability in the employees and the facility which will manufacture the product.

Technical feasibility is of real concern in such projects as putting a man on the moon, laser communications, a hydroelectric power project, and a host of technically oriented design problems.

In some cases, like government sponsored projects, the economic feasibility appears to be of no concern. This is a fallacy. *Good* design practice ensures that the resource exchange is in proportion to the needs being met. By comparison with other ways of meeting the need, a measure of economic feasibility may be obtained (also called cost-effectives or cost-benefit analysis).

A CASE HISTORY OF TOYS
FOR BLIND CHILDREN

A design team was working on a toy which would appeal to blind children. The following is an excerpt from the report of one of their meetings. They were at the stage of considering several alternative approaches to a construction toy.

"Sketches of assembled models were submitted and the discussion centered around them. The blind child is no different from the child with sight in that he will also build cars, boats, houses, etc. From these basic shapes, discussion centered on the structural members and in particular, the joints. (Because the users were blind, the Human Factor Feasibility would be all right if the structural members had different cross-sectional shapres for identification by feel.)

Comments: It was the consensus of the group that the tooling costs involved in such a system could be prohibitive. Therefore, the suggestion was that the main members should be an established product (e.g. plastic extrusions, conduit, metal tubing) . . . The main consideration was to make the components from inexpensive materials and processes."

The design team decided not to use expensive tooling. This arbitrary decision became in effect a crucial criterion. It tended to eliminate solutions involving any tooling at all. Later they found out that some types of tooling are not expensive. Ingenious schemes are always being devised to reduce tooling costs on low-volume products. For example, low-volume automobile bodies may use fiberglass on wooden forms. On the other hand, in large quantities, automobile bodies are shaped and cut from metal by expensive dies.

One can well imagine that the market for toys for blind children is relatively small, and one would not be likely to use injection molding of plastic parts for a few hundred sets of the construction toy in question. Yet, on the other hand, if you have a good idea, the world may be your market. For a cause like this, we should not overlook the financial assistance that might come from government or private sources. Resource feasibility is a good test to apply, but occasionally, we design and make things which do not need an economic rationale.

GUIDELINES TO SCREEN FOR FEASIBILITY.

Do	Don't
Do ask yourself if it will work under the circumstances that prevail.	*Don't assume that your great idea for a technical solution will be successful in every situation.*
Do consider the financial and human resources needed to get the alternative implemented.	*Don't be trapped into thinking that there will always be funds and expertise when they are really needed.*
Do consider the time it will really take.	*Don't make our reputation any worse than it is. Engineers tend to be overly optimistic on how long a job will take.*
Do consider the people who physically come into contact with the engineered design or with its undesirable outputs.	*Don't assume that if you can manage the controls, that a less enthused operator will do the same.*
Do think about the impact on society. You will be glad that you did.	*Don't leave it up to others to smooth the feathers of an angry social group. Another design solution could blunt the impact.*

ASSIGNMENTS TO ASSIST YOU TO GET BETTER BY REVIEW AND PRACTICE

Questions On the Content of This Chapter

1. Examine the Checklist for Technical Feasibility. Which of these would be important in the design of a dust precipitator which would be mounted inside an existing chimney?
2. Examine the Checklist for Resource Feasibility. Which of these would be important if the above precipitator was being designed by the technical staff of a large and profitable mining company?
3. Would Human Factor Feasibility be important in the design of the above dust precipitator?
4. Would the Social-cultural Feasibility be important in the design of the above dust precipitator?
5. In this chapter, the resource feasibility was shown for a TV set with a production quantity of 1000 units. What happens to the cash required and to the selling price if 10,000 units are used for the calculations?

Question for Review of the Case History

Clearly the human factors of toys for blind children are important. Would the Social-Cultural Feasibility be likely to be important too?

ASSIGNMENTS FOR PRACTICE

1. If you are following through on these chapters with your practice project or a real project, you should have a list of alternatives, including a few which are not clearly feasible. Screen the list for feasibility, using one of the models shown as applications in this chapter.
2. Take a look at some completed things like equipment, products or large-scale capital projects. Do you think that all of the feasibility tests were properly considered? Do this a few times and you will improve your alertness in screening for feasibility.

8 How to Decide on the Best Design Alternative

KEY IDEA: THE BEST CHOICE OF DESIGN ALTERNATIVE IS THE ONE THAT WOULD ACHIEVE THE MOST OVERALL SATISFACTION OF THE DESIGN OBJECTIVES.

HOW WE TOOK THE EMOTION OUT OF A DESIGN DECISION

When I was a manager of engineering design, one of my early design crises was due to the concern over a single design parameter: the maximum line voltage for life-testing a television set. If the TV designer accommodates 140 volts for life-testing, then the design-center voltage is moved from 115 volts to 125 volts. Consequently, the performance at ninety volts is not so good. At the low voltage extreme the picture is not wide enough to cover the television screen. If the design-center voltage is not changed upwards, then the life-test failures at 140 volts are excessive.

The manager of quality control could not agree with my team of television engineers on the extremes of line voltage for which satisfactory performance was required of a television set. (Our real problem was that we did not know how to design a television set that could meet these two extremes, but we did not admit it, even to ourselves.) Our position was that if the TV set was labeled for operation between 105 volts and 130 volts, then the matter was

147

taken care of. The quality control manager, who was a person who couldn't be pushed around, insisted that he had measured 140 volts at one time at his house, and so that was the high voltage he life-tested them at.

You and I know that an engineered design cannot be expected to function at some extreme of use unless it is designed for that in the first place. These TV sets were designed to function in the range of 100 to 130 volts and anyone who tested them at 140 volts and insisted on good operation at 90 volts also, was doing so at his/her own peril. Nevertheless, this position of hardened attitudes led to one crisis after another. A sample batch of TV sets would fail the life-test and the factory would be required to rework all of that batch.

Frequently, production was shut down while waiting for better quality components. Our engineers worked overtime trying to solve the multitude of problems that were being generated. The sales manager listened to these positions and shook his head in frustration because he didn't know who to believe—and he wasn't getting enough production to fill his orders. Intermittent delivery of TV sets was the order of the day and the problem-solving meetings sometimes wasted days and evenings.

How could this impasse be solved? What was needed were some cold hard facts about the line voltage to which the TV sets were being subjected. A number of calibrated voltmeters were obtained and sent out to servicemen all across the country. They were asked to measure the line voltage at the time that they serviced a TV set. They were asked to take some early morning and late evening readings as well.

When the data rolled in the problem was not immediately solved because there was solid evidence of extremes of line voltage. By using intuition alone the results could be interpreted to support any position. To make a reasonable decision about the design parameter of line voltage we constructed a decision matrix like the one shown in Figure 8.1.

The two alternative designs being considered were:

1. The present one, having a design-center voltage of 115 volts.
2. The present one, modified by a transformer having a design-center voltage of 125 volts.

VOLTAGE INTERVAL	85–95	95–105	105–115	115–125	125–135	135–145	EXPECTED AVERAGE HOURS
% OF CUSTOMERS WITH VOLTAGE	5	10	30	40	10	5	
ALTERNATIVE 1 (Current design)	1500 (U)	1000	500	300	200	50	437
ALTERNATIVE 2 (Modified transformer)	2000 (U)	1500 (U)	1000	500	300	200	790

U = UNSATISFACTORY PICTURE WIDTH

Entries are the average number of hours that a TV set would be expected to perform at the center point of the voltage interval.

Figure 8.1. A Single Parameter Decision Matrix.

(A mechanical engineer may think of this example as being two operating temperatures of a machine. A civil engineer may find it comparable to life-testing a roofing material.)

The alternative design number 2 would result in about 10 percent less voltage being applied to the vacuum tubes and considerably less heat as well. However, as indicated by the U's in the first two columns, such a design change was not fully satisfactory because the picture did not cover the width of the screen. Nevertheless, the life was extended considerably. The entries in the matrix are the average number of hours that the TV set could be expected to perform at the center point of the specified voltage interval for the column.

On the right hand side is the expected average hours as determined by the distribution of customer voltages. For example, the first row sums up as:

5% × 1500 hours + 10% × 1000 hours + 30% × 500 hours
+ 40% × 300 hours + 10% × 200 hours + 5% × 50 hours
= 437 hours, on the average.

The average hours (weighted sum) for the second design alternative is 790.

Clearly, the second design alternative is superior for the expected hours of use, except that 15 percent of the customers would experience a picture which would not fill the screen.

This matter was discussed with the sales department and they felt that they could live with the lesser problem and pay for a minor field modification whenever required by a customer having extremely low line voltage. Thus alternative number 2 was chosen. As well, the quality control manager was persuaded to back off to 135 volts for life-testing.

Two good things resulted from having the facts. Firstly, *most of the emotion went out of the decision-making.* The meetings that followed were very effective in problem-solving because the hostility in the design decision had been defused. The second good thing *was that we had data to use for future designs.* Our engineers had to believe, like to or not, that television sets did in fact experience extremes of 85 volts or 145 volts in about 10 percent of the cases. A really good design would look after these extremes. The happy ending to this design situation was that once we were relieved of the day-to-day production problems arising out of excessive line voltage testing, we got down to the job of making a really good design which worked beautifully at all of these extremes.

WHAT THIS CHAPTER WILL DO FOR YOU

As you work your way from concept phase to the implementation phase, there are many places for serious decision-making. The major place is at the end of each phase. You evaluate the work done and assess the chances of reaching target objectives. You decide to go ahead, modify targets or abort the project.

Within each phase, the important decision point is at step 6 of the systematic design procedure, called "Select the solution." Prior to this you should have done "Create Alternatives" and "Screen for Feasibility." You should then have a few feasible alternatives, and from them you must select the one most likely to satisfy the objectives and specifications which were developed earlier.

The objective of this chapter is to show you *how to make a quantitative design decision with the use of a judgmental scale.*

The benefit to you of making your major design decisions this way is that you will get *better decisions.* You will also reduce the effect of any personal bias that might get in the way of your success in making the right decisions.

By reading about my own experience in decision-making at the beginning of this chapter, you can see that there are many cases where a systematic decision will defuse the emotional fireworks we get with intuitive decisions. We will examine the real differences in decision-making styles. Then we will develop the **Principle of Satisfaction of Objectives**, which is the rationale for good design decisions.

You will learn how to use a decision matrix for intuitive-systematic decisions and thus get the best of both worlds—experience and new facts.

The examples in this chapter reduce theory to practice for the case of a consumer product and the case of a large-scale capital project. The commentary shows you how to review and revise the decision matrix until you are sure of the answer you get.

The Guidelines give you a compact set of reminders for good decision-making.

At the end of the chapter are questions to stimulate your mind in a review of the concepts. There are suggested things for you to do on the job so that you can become recognized as a rational decision-maker.

KINDS OF DECISIONS

Intuitive Decisions

These are the decisions that people say are based on "gut feeling."

Like decisions in love, intuitive decisions are sometimes good and sometimes bad. In fact, some of the best and some of the worst decisions in history were made intuitively.

A purely intuitive decision is primarily a property of the mind rather than a property of the object about which a decision is made.

Systematic Decisions

If the properties of the design object are measured and used to make a decision between design alternatives, this is a systematic way of reaching a decision. It can be repeated and reviewed by any other design authority. It can be audited. It is an "objective" decision. A decision to choose an engine by the measured efficiency would be an example of a systematic decision which can be objectively reviewed by other persons.

Expert-Intuitive Decisions

A person making an intuitive decision may have much expertise on the subject matter and really be basing it on a lot of collective experience which is not readily explicable. In other words, when the expert makes a decision within his field of knowledge, it may still be based upon intuition and "gut feeling." This is very much different from a person having the authority to make a design decision—not knowing what it's all about—and insisting on having his or her own way. That can be a disaster.

Intuitive-Systematic Decisions

Many design decisions cannot be based on the measured performance parameters alone. There are many intangibles that must be evaluated as well—esthetics and convenience among them. Intuitive judgments can be converted to numercial scales in a systematic way. Then, the decision can be reviewed and revised, as well as be explained. We will regard an intuitive-systematic decision as being both rational and objective. This type of decision will be described in the following section.

THE PRINCIPLE OF SATISFACTION OF OBJECTIVES

Real engineered designs are made for a purpose. The purpose or purposes are properly described by a set of objectives which follow from a needs analysis, as was explained in chapters two and three. The objectives should state what is to be achieved by the design solution. It is the satisfaction of these objectives that we are going to measure.

As explained in chapter two, the new state of affairs brought about by an engineered design can be expressed by a state vector, S_2, which has numerical values assigned to all of the design variables and affected environmental variables. This is expressed by:

$$S_2 = \overline{/X1, X2, X3 \ldots Y_1\ Y_2\ Y_3 \ldots Z_1\ Z_2\ Z_3\ \ldots}$$

You may recognize this as a set of specifications and drawing dimensions. Indeed, a large set of them goes a long way to describing the new situation. However, it must be remembered that it is the objectives which we are trying to satisfy by means of the specifications and drawings. Because there is usually more than one objective, the variables of S_2 can be grouped into subsets of variables which are in a new state in order to achieve satisfaction of the multiple objectives. In other words, we may assign the X variables to the satisfaction of objective X, and the Y variables to the satisfaction of objective Y.

THE PRINCIPLE OF SATISFACTION OF OBJECTIVES.

The design variables are set at values which achieve a high over-all satisfaction of the design objectives.

So much for the theory. What we really need is some numerical measure of the satisfaction of objectives and this should be done in some "objective" fashion so that the decision is credible, reviewable, and good.

Since a state vector, S_2, would normally contain thousands or even millions of variables, it is convenient and advantageous to express the new state as the satisfaction of a set of objectives, O_x, O_y, ... These objectives will say what is to be achieved by the performance aspect of the design.

We will also have objectives pertaining to the timing of the project and to the cost of the resources to be used. As any experienced designer knows very well, the performance, time and cost objectives are very much interrelated when a design is being created. There may be tradeoffs among technical objectives and there will certainly be tradeoffs between performance, cost and time. The way to take these tradeoffs into account is to use the decision matrix which will be explained below.

THE PERCENT SATISFACTION
OF AN OBJECTIVE

In order to make a systematic decision among design alternatives, we must have common units for comparison. This unit is the Percent Satisfaction of Objectives. It can be weighted and summed into a figure which gives the Overall Percent Satisfaction of Objectives.

Let X_i be the percent satisfaction of objective O_i which results from a subset of design variables I_1 I_2 I_3 . . . To assign X_i a value requires you to quantify your knowledge and intuition about the satisfaction gained from some part of a design. This is all right if you have the expertise to assign a value to X_i.

Less than 50 percent satisfaction of any single objective may put a design in the unsatisfactory region where it cannot be compensated for by additional satisfactions elsewhere. For example, if you only achieve 30 percent satisfaction of the objectives regarding seating capacity of an automobile, and you get 90 percent satisfaction of the speed objectives, it will not do to average these out to 60 percent because one may be below the threshold and therefore unsatisfactory.

To take a weighted sum of satisfactions we will use the linear equation which is:

$$\text{Overall \% Satisfaction of Objectives} = \Sigma w X_i$$

where:

w_i is the weight given to objective O_i

X_i is the percentage satisfaction of objective O_i

The sum of the weights is equal to 1.0.

If we use this linear equation, then we should have some reason to believe that the quantities, X_i, are also linear. This is usually the case over most of the range of satisfaction.

Consider the S-curve shown in Figure 8.2. Let it represent the percent satisfaction of the speed of the vehicle. For the use which the vehicle is to be put, there is a minimum level of speed which gives a threshold value of satisfaction. Beyond a certain speed, we get into saturation effects.

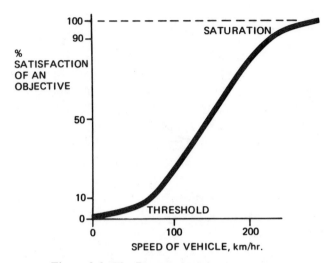

Figure 8.2. The Percent Satisfaction Curve.

This S-curve is based on the physical and psychological satisfactions to be derived from almost any design parameter. Since
sensation is proportional to the log of the stimulus, the S-curve
transforms to a reasonably straight line over most of its useful region.
This is shown in Figures 8.3 and 8.4. The only region to avoid is
that of satisfactions below the threshold value.

This Percent Satisfaction scale has been used with great success in
hundreds of designs with which my consulting firm has been associated. The extent of agreement which is possible among engineers and users as to the percent satisfaction of an objective is quite
surprising. Usually a small amount of discussion and a few revisions
will result in an agreed upon value. If not, there is a special numerical
consensus technique which will be described in Appendix II.

So much for the Percent Satisfaction Scale. Let's go back to the
expression:

$$\text{Overall \% Satisfaction of Objectives} = \Sigma w_i X_i$$

This can be represented by a row of the Quantitative Decision
Matrix shown in Figure 8.5. For simplification, I have shown a
case where a decision is required on a technical basis only. This is

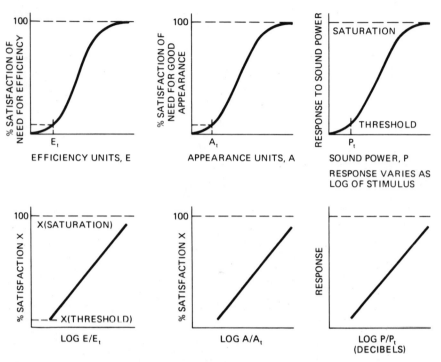

Figure 8.3. Converting Parameter Scales to a Common Percent Satisfaction Scale.

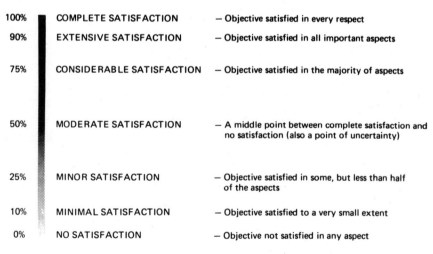

100%	COMPLETE SATISFACTION	— Objective satisfied in every respect
90%	EXTENSIVE SATISFACTION	— Objective satisfied in all important aspects
75%	CONSIDERABLE SATISFACTION	— Objective satisfied in the majority of aspects
50%	MODERATE SATISFACTION	— A middle point between complete satisfaction and no satisfaction (also a point of uncertainty)
25%	MINOR SATISFACTION	— Objective satisfied in some, but less than half of the aspects
10%	MINIMAL SATISFACTION	— Objective satisfied to a very small extent
0%	NO SATISFACTION	— Objective not satisfied in any aspect

Figure 8.4. Percent Satisfaction of an Objective.

OBJECTIVES WEIGHTS, w ALTERNATIVES	P1 SEATING SPACE	P2 ACCELERATION	P3 FUEL CONSUMPTION	P4 APPEARANCE	OVERALL % SATISFACTION OF OBJECTIVES	INTUITIVE RATING ON THE WHOLE
	.2	.3	.3	.2	ΣwX	
VEHICLE ALTERNATIVE A1	95%	80%	70%	90%	82%	90%
VEHICLE ALTERNATIVE A2	75%	80%	60%	80%	73%	70%
VEHICLE ALTERNATIVE A3	60%	40%	80%	50%	58%	50%

Entries, X, are the percent satisfaction of technical objectives. This matrix assumes that the time and cost objectives are otherwise satisfied.

Figure 8.5. Quantitative Decision Matrix.

required when the cost and time objectives can be met with any of the alternatives being considered. (The application examples shown further on in this chapter show how to take time and cost into account also.)

Note the column on the right which is labeled "Intuitive Rating on the Whole." This is a check on the completeness of your weighted objectives. Without considering the parts of the decision, think of each alternative as a whole and rate them by that "gut feeling." If another alternative scores highest, then either your intuition is wrong—or—you have missed something. In the final decision matrix, your systematic decision and your intuitive one should be the same.

APPLICATION TO A CONSUMER PRODUCT

Electric Toothbrush: Selection of the Design Solution

Let us assume that a "small" company has three choices that appear feasible. Below is a decision matrix which uses the design objectives which were established near the beginning of the design process. The outcome entries in the matrix, X_i, will be the percent satisfaction of a design objective, O_i.

Assuming we are satisfied with the values in the decision matrix of Figure 8.6, we see that the first two choices dominate the third. However, there is not enough difference in the scoring to differentiate between the first two.

Actually, we have missed out on a very important performance objective: Psychological concern for safety. (An intuitive rating on the whole may have led us to this conclusion.) Even though UL may give approval to a line operated electric toothbrush (as they do for electric razors), many people would not want to risk using it. Thus, if this objective were added, the first alternative would score higher.

Having constructed a decision matrix, the designers now have a way of reviewing and revising their decision until no further improvement can be made. They also have a way of scoring improvements as they optimize their selection at a later design phase.

DESIGN OBJECTIVE WEIGHTS, w	O_P ATTRACTIVENESS RATING BY CONSUMER PANEL	O_P TECHNICAL FUNCTIONS RATED BY DENTAL CONSULTANT	O_C SELLING PRICE TARGET ACHIEVEMENT	O_C COST OF MEETING ADDITIONAL STANDARDS	O_T TIME TO PRODUCTION	ΣwX OVERALL % SATISFACTION OF OBJECTIVES
ALTERNATIVE DESIGN	.4	.2	.2	.1	.1	
RECHARGABLE BATTERY; BRUSH MOVES SIDEWAYS	70% / 28	90% / 18	50% / 10	60% / 6	90% / 9	71%
LINE POWERED; BRUSH MOVES SIDEWAYS	80	90	80	20	50	73
REPLACEABLE BATTERY; ROTARY RUBBER	40	50	100	100	40	60

O_P = PERFORMANCE OBJECTIVES
O_C = COST OBJECTIVES
O_T = TIME OBJECTIVES

Figure 8.6. Example of Use of the Percent Satisfaction Decision Matrix for a Consumer Product.

APPLICATION TO A LARGE-SCALE
CAPITAL PROJECT

A Hydroelectric Dam for Latin America:
Selection of the Design Solution

Suppose that the number of feasible concepts is only one, and that the crucial decision is the site for the dam.

The criteria and their weights which are now to be used for evaluation, were developed prior to solution finding (chapter four). For this design project, most of the important criteria are technical and resemble the objectives. In the actual selection of a design solution for a project as large as this one, a larger set of criteria would be used. This example is just an overview.

DESIGN OBJECTIVE OR CRITERION / WEIGHTS, w / ALTERNATIVE DESIGN	O_P POWER 4 ± 1 MW	O_P EXPANSION OF POWER	O_P IRRIGATION WATER	O_P WATER TRANSPORT	O_P OBJECT OF NATIONAL PRIDE	O_C CAPITAL COST	O_C POWER COST	O_T TIME TO COMPLETION	ΣwX = OVERALL % SATISFACTION OF OBJECTIVES
	.3	.1	.2	.05	.05	.1	.1	.1	
EARTH DAM, SIDE CHANNEL, GRAVITY IRRIGATION, SITE A	90 / 27	100 / 10	60 / 12	50 / 2.5	75 / 3.7	60 / 6	70 / 7	50 / 5	73
AS ABOVE, SITE B	75 / 22.5	100 / 10	60 / 12	50 / 2.5	50 / 2.5	70 / 7	70 / 7	75 / 7.5	71
AS ABOVE, SITE C	70 / 21	70 / 7	100 / 20	100 / 5	100 / 5	100 / 10	90 / 9	100 / 10	87
	X / wX								

O_p = PERFORMANCE OBJECTIVES
O_C = COST OBJECTIVES
O_T = TIME OBJECTIVES

SENSITIVITY OF THE DECISION: By looking at the products of weights and percent satisfaction in the lower right corner of the boxes in the matrix, it can be seen that Site C gained the decision mainly by satisfying the objectives for irrigation water and capital cost. If the weights of these are reduced and the weight for power increased, the decision could be switched to Site A. However, some optimization of the design on Site C could put it clearly in the lead.

Figure 8.7. Design Matrix for a Large-scale Capital Project.

PROBABILITIES OF SUCCESS

There is a body of knowledge called Decision Theory, which uses subjective probabilities for events that would affect the outcome of a decision, and then uses probable payoff analysis. If you wish to go deeply into this subject, I recommend a book written by an associate of mine, Dr. M. W. Lifson, *Decision and Risk Analysis for Practicing Engineers,* Cahners Books, 1972.

What follows is a summary of a decision analysis I did for a client on which way to go on the development of a highly specialized and expensive test system. The details have been modified so that the client will not be revealed. However, this has a benefit. In generalizing the entries, I have made it more useful for other situations.

By use of a decision tree, you can chart out the potential consequences of decision alternatives. By a "roll-back" technique you can then calculate the course of action which is best, provided that you are willing to gamble on averaging out your wins and losses. An example of a speculative business decision will show you what I mean before we go to a sophisticated engineering decision.

Suppose solutions to a problem of maximizing production involve a choice of making rainwear or sunsuits, and suppose also that the probabilities of wet and dry seasons are .8 and .2 (based on weather records). A decision to choose either solution will be followed by a chance event of weather as shown in Figure 8.8. The outcomes in this case are a profit or a loss in dollars.

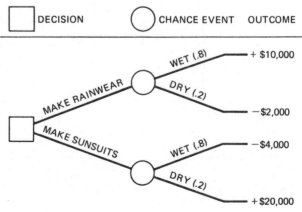

Figure 8.8. A Decision Tree.

When outcomes are multiplied by probabilities, the expected monetary value of each solution is worked out. For rainwear, it is .8 × $10,000 + .2 × (−$2000) = $7,600. For sunsuits, it is .8 (−$4000) + .2 × ($20,000) = $800. Clearly, rainwear is ahead so far as this example goes.

Now on to a serious type of decision for an engineered design. Suppose there is a pressure to get the thing done in a hurry. Should we jump right in and do the final design or should we do a design iteration with some pilot tests first? What are the risks? The next example is abstracted from a detailed decision analysis I did for a large corporate client. It took 20 man-days of our time to do the decision analysis.

In Figure 8.9, you see two major forks of the decision tree starting at the decision box on the left hand side. If we follow the top fork we will spend $60,000 for pilot tests on subsystems. That is called the "toll cost." After doing the tests we will have some new informa-

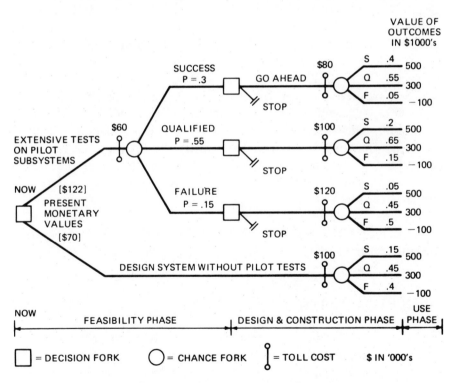

Figure 8.9. A Design Decision About an Extra Iteration.

tion on which to base our next decision. The probability of a "successful" outcome was assessed by the project manager to be .30;
of a "qualified success" to be .55, and of a "failure" to be .15. These
were subjective probabilities, based on the best knowledge available
at the time of the decision analysis.

Let us follow this reasoning a bit further into the future by following the top branches of the decision tree from left to right. If the
pilot tests were to be successful, the project manager would decide
to go ahead. They would then spend $80,000 on construction of the
full system (another toll charge). The project manager estimated
that the probability of full success with the completed system would
then be .4, and if so the outcome would be worth $500,000. (The
values assigned to outcomes were obtained by a group assessment.)

Next we "roll-back" the future outcomes to the present point by
making calculations from right to left. First, multiply the outcomes
times probabilities at the last chance event on the right. We get an
equivalent monetary value of .4 × $500,000 + .55 × $300,000
+ .05 × $100,000, which is $360,000. However, this has a toll cost
of $80,000, so the value at the right hand decision point is $360,000
− $80,000 = $260,000. At the next chance event to the left we
obtain .3 × $260,000 + .55 × (something from the middle branch)
+ .15 × (something from the lower branch), all less a toll charge of
$60,000. We would then arrive at the present monetary value of
$122,000. This is the value, on the average, if we made the decision
to go by the top fork of the decision tree (pilot), not just once but
many times. By the same procedure, we can show that the present
monetary value of the lower fork is only $70,000. If we are a ra-
tional decision maker, we would prefer the top fork because the
other one would average out to much less. In this case the decision
was to spend the $60,000 on extensive tests on pilot subsystems
because on the average this action would pay off the best.

WHEN TO USE DECISION TREE ANALYSIS

What is the real value of a decision tree analysis? In my experience,
when we applied them to decisions worth $50,000 and up, the
analysis was worth it. Once I charged a client $5000 and saved
them a million dollars by steering them off a design path that was

being pushed too hard by an uninformed executive. In the example explained above, we saved $125,000 by discovering that one could borrow an expensive computer instead of buying it. That was tangential to the major thrust of our analysis but it makes my point. *The major value of this kind of analysis has been to closely examine the decision problem,* to think it through, to assess probabilities and outcomes, to explore old and new avenues, etc. So learn to do it on scraps of paper and reap 90 percent of the benefits. When you discover a really obtuse decision problem, you can still call in a decision analyst to pin it down, to simplify it, or to run it off on a computer.

THE BEST DESIGN IS KNOWN
ONLY BY COMPARISON

You could spend a lifetime looking at alternatives and never be sure that you have the very best that is possible. There are always restraints of time and money on a design project, but our problem is solved when we come up with something which is reasonably close to the best. This is only known by making a comparison of a number of good designs. Use the quantitative decision matrix shown earlier in this chapter. When you choose the one with the highest percent satisfaction of objectives, be sure that this satisfaction is somewhere in the upper quartile or you should have some explaining to do to yourself. More about this in chapter ten on optimization. At this point you should be able to make a systematic-intuitive decision among a number of viable design alternatives which have passed the feasibility test. In the next chapter we will deal with the implementation and communication of the design solution.

SUCCESSFUL DECISION MAKING

In this chapter, I have shown you how to make a major design decision. With a decision matrix you make the best of both worlds: the engineering world of measurable design outputs, and the intuitive world of your life experience. Use it until it becomes second nature to you, then much of the approach will remain with you when you make those thousands of minor design decisions which can be made quickly and without quantitative analysis.

The decision matrix is great for nonengineering decisions too. My associates and I have adapted it for use in business and government decisions. You can do the same when you have had practice. Choose some exercises from the end of this chapter and put into practice what you have just read.

DESIGN GUIDELINES

GUIDELINES FOR DESIGN DECISIONS

Do	Don't
Do get the facts before taking sides in a technical issue.	*Don't get hot and bothered when your ego gets bruised by decisions made irrationally by others.*
Do remember that your opinion has value only when you have a corresponding expertise.	*Don't feel compelled to sound off about decisions made by others when the matter is outside your area of expertise.*
Do focus your efforts on the balanced satisfaction of a set of design objectives.	*Don't get carried away with technical sophistication unless it contributes to the design objectives.*
Do set up a quantitative decision matrix for major design decisions.	*Don't feel that a fast decision is always best. Don't belittle the time and effort required for a decision analysis.*
Do choose a decision model from this chapter and apply it to your own design work.	*Don't just read this chapter and expect to become a great decision maker.*
Do a double check on a matrix by comparing it with your intuitive rating of the whole.	*Don't fall into the trap of believing in numbers until you are sure of them.*
Do recognize that some design decisions are so uncertain that you should use probabilities.	*Don't construct a decision tree with so many branches that you get lost in the bushes.*

Questions to Help You Consolidate Your Understand of This Chapter

1. When is an intuitive decision likely to be a good one?
2. How would you describe a rational decision?
3. Why should design decisions be based on the overall satisfaction of design objectives?
4. Under what condition is the "percent satisfaction of an objective" scale reasonably linear?
5. How are cost and time factors brought into a major design decision?
6. When is a major design decision amenable to analysis with a decision tree?

Questions of the Case Studies

1. Consider the single parameter matrix of Figure 8.1. Suppose a third alternative design is to be considered in which the design center voltage is changed to 135 volts. Estimate the expected average hours for this alternative. What reasons would you have for recommending for or against this alternative?
2. Examine the decision tree of Figure 8.9. How much could be invested in the pilot tests, and still have the same present monetary value as for the decision to go ahead without the pilot tests?

Assignments to Help You Become a Pro

1. If you are praticing your skill in a minor design project as suggested in chapter one, make a quantitative decision matrix for it and show how you made the decision among the major design alternatives that have passed the "screening for feasibility."
2. Refer to the Application to a Large-scale Capital Project. Suppose that the objectives be changed so that the major purpose of the dam is to obtain irrigation water. Increase its weight to .4 and reduce the weight on power to .1. Recalculate the

overall percent satisfaction of objectives for each site. Which one is in the lead now?

3. Do a post-mortem on an engineered design that you did, or with which you were associated. What were the major decisions that were made or should have been made? List them.

   For each decision, classify it as being made mainly intuitively or mainly by a systematic method.

   Construct a quantitative decision matrix using known or supposed design objectives. Weight them. For rows, use at least two design alternatives, including the one that was used. Enter the percent satisfactions and obtain the weighted sums.

   Most likely your analysis will confirm the decisions made at the time. Occasionally, there is only one way to go and everybody knows it. However, since hindsight is better than foresight, you may even show that there could have been a better decision.

   The main purpose of this exercise is to get you familiar with the decision matrix by using it on familiar ground.

4. Choose a quantitative decision model from this chapter and apply it to a real design project as soon as you can. Work out the bugs in private, discuss it with a colleague, and then present it to your boss or client for comment.

9 Effective Communication Makes a Successful Transfer

KEY IDEA: YOU CAN IMPROVE THE TRANSFER OF THE DETAILS AND INTENT OF AN ENGINEERED DESIGN BY APPLYING SEVEN RULES OF DESIGN COMMUNICATION.

DON'T BREAK THE UMBILICAL CORD TOO SOON

A California manufacturer of TV sets decided to move its manufacturing facility to nearby Mexico. The rationale was that the savings in labor would be substantial. The corporate offices, the marketing and the engineering would remain in California. There were precedents for such a move in that several large electronic manufacturers had moved their assembly plants to Taiwan, Hong Kong, Korea, etc. It seemed logical to the president of the California company that drawings and prototypes could be shipped any distance and that the TV sets could be manufactured and assembled according to the specifications.

It took about two years for the experiment to run its course. The initial problems of shaking down the new manufacturing facility were expected and chalked up as "growing pains." However, profitability was never realized. The experiment was terminated by the company going out of the TV business altogether. Unfortunately, for the design team, their jobs also went down the drain.

This company had a good track record for innovation, its engineers were clever. For years they had issued drawings and specifications to the manufacturing unit who officially carried them out after the sign-off. Unofficially, there was still a continual "mothering" of the manufacturing unit to help them through with the intricacies of innovations in the product. The engineers were constantly in touch with manufacturing, quality control and purchasing. When something unusual developed, personal communications supplemented the specifications. The "umbilical cord" with mother engineering was still functioning.

What happened with the Mexico experiment was that the ordinary communications difficulty with innovative engineering design was exaggerated by the additional dimensions of distance and language. The umbilical cord was cut at sign-off time. The project simply could not stand the strain. It fell apart.

THIS CHAPTER HAS MANY USES

Communicating the design details and intent is not an easy matter. What was learned over thousands of hours must be transferred in a relatively short period. If you learn to do this transfer well there will be spinoffs for you. You will do better in all forms of communications.

Objectives

1. *You will learn about the Principle of Design Communication.*
2. *You will learn seven Guidelines of Design Communication.*

Benefits

1. *You will be able to reduce the time of post-design problem solving by communicating the design details effectively in the first place.*
2. *You will do better with an engineered design because the results will be as you intended.*
3. *You will be able to communicate better with persons within and outside your organization.*
4. *You will be able to successfully introduce a major technical innovation to your organization.*

The step, "Communicate the Design Solution" occurs at the end of any phase. It is particularly important if a new team takes over for the next phase. Therefore, it is most important when the detail design is done because nearly always someone else takes it over to make it.

In this chapter, you have seen how a manufacturer cut off his own lines of communication with the design team. I alluded to the basic reason for communication failures with engineered designs. This leads to the Principle of Design Communication which is stated in a way to make it easy to remember. There are seven important guidelines of effective design communication. These are explained with examples and summed up in Guidelines for Design Communication.

An application to an Equipment Design shows how to plan an effective transfer for a mainly technical product. Next is an Application to a Consumer Product which is a little more difficult because of the intangibles involved. In the Application to a Capital Project, you have an example of a communication plan for transfer from concept phase to a feasibility phase.

The case history of a misinterpretation of a drawing change order demonstrates how difficult it is to communicate an idea in detail.

There are questions and assignments for you to follow-up to become more knowledgeable in design communication.

THE BASIC PROBLEM OF COMMUNICATING
THE DESIGN

Any written or verbal communication about a real object involves a tremendous reduction of information. Take an automobile design, for example. The engineers could talk about the details of a prototype for several months and still not get "talked out." Yet, it is common practice to communicate a design like this to other persons of other disciplines with orthographic drawings, written instructions and some samples. The fact that the communication works at all is amazing, considering that the information reduction must be at least a million to one.

Consider the communication of the design intent of a fender. Orthographic drawings and specifications help pin down some of the critical parameters. Photographs and samples go a long way too, because a picture is really worth a thousand words. The real thing is worth a million words. The reason why it is possible to live with a high information reduction on a part like a fender, is that recipients already have a good understanding of what the communication is intended to mean. This is not true of something radically new in a design which is to be communicated. We have a real problem in communication. It helps to know something about the basics of communication so that we can be as effective in the difficult areas as in the tried and proven ones.

In addition to the tremendous information reduction, whatever is communicated is often distorted by words which are supposed to represent things, distorted by the jargon of specialists, and distorted by the technical distance between the engineers who design and the workmen who put it together.

Having the design team building it and using it would be the best solution to communicating the design intent—but this is seldom practicable. It is more practical to use some time-tested rules of communication, and to phase in the communication over as long a period as you can. In the rest of the chapter, you will learn a set of guidelines which will make the communication of the design intent effective. They support the major principle of design communication.

THE PRINCIPLE OF DESIGN COMMUNICATION.

The what, *the* how *and the* when *of design communication should be determined by the* who.

That is, all communication should be oriented to the recipients who will implement the design solution.

GUIDELINE NUMBER 1: MAKE IT
UNDERSTANDABLE TO THE RECIPIENT

The most important thing about design communication is that the message must be understood. You cannot "make people understand" your communication—but you can "make it understandable." *The responsibility for success of design communication is clearly on the sender.*

No matter how clear and simple a design communication may seem to you, others often have a different understanding of what you mean. For instance, people without engineering training cannot be expected to understand technical subjects in the same way you do. Technical terms and drawings which make it possible for the engineer to communicate effectively with his technical associates, can be a barrier to communication with those who make, sell or use the finished design. *Technical jargon stands in the way of effective communication with anyone but your technical peers.* The specification for testing a piece of electronic equipment may read as follows: "Increase the positive bias on the CRT until the luminous output is 50 lumens:. Equivalent instructions to a nontechnical tester in the factory should be, "Turn up the brightness control until the brightness meter reads 50."

Besides technical jargon, you must avoid the "argot" which is part of the language of a private work group. I can recall some TV engineers talking about the "frying sound" from a TV circuit and knowing clearly what was meant. This was not even technical jargon that could be communicated to other technical persons. It was argot, only good for team-mates. Household or work-group words get so familiar to us that we often fail to realize that they are not understood by others. Yet they find their way into our written communications.

There is one language that is known to all technical disciplines and to nontechnical people alike, and that is plain English. Blessed is the engineer who uses plain English as often as he can. We never have to call him in and tell him K. I. S. S. (Keep it simple stupid).

Keeping it simple is a good rule to apply when communicating upwards to management too, but there is an additional point. The information requires further reduction, not by condensing the design details, but by picking out the most relevant parts and communicating them directly. For instance, there may be a drawer full of drawings about a new factory which is to be built, but the important communication to management is what it will do, how much it will cost, and when the money is needed.

GUIDELINE NUMBER 2: ARRANGE IT IN
THE ORDER REQUIRED BY THE USER

The second most important thing is to arrange the communication in a sensible sequence. In communicating with the advertising department, for instance, it is better to list features in order of customer preference rather than in the importance you give to the technical features. In the design of a hand mixer, being light in weight and easily cleaned is more important than the reinforced handle and the ease of repair.

The method of large-scale production is one of the first things the production group will have to know. In writing test instructions, the order they are done in the factory is probably different from the order they were done in the laboratory. The test instructions for the service manual will be in a different order again because the product is taken apart instead of put together. Unfortunately, most of our service manuals are written by the design engineers and whether the design object is a radio or a hydrofoil boat, a designer thinks about the order of putting it together, whereas the maintenance personnel thinks of the order of taking it apart.

The reasonable approach to developing a sequence according to the user is: Imagine yourself as a user of the instructions or specifications and ask yourself, "What is the first thing I would want to know? What is next? And what is the last thing I would want to know?"

GUIDELINE NUMBER 3: CHOOSE THE MOST
EFFECTIVE METHOD OF COMMUNICATION

Knowing that effective communication is difficult at best, take a moment to consider the most effective method of communicating the design in every situation. Not everyone understands the orthographic projections used on blueprints, and even if they do, perspective drawings are an aid to understanding. In other words, depending on who the recipient is, you can choose a method of communication which is most effective in that situation (See Figure 9.1). Below is a checklist of the most frequent methods of communication.

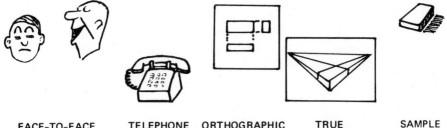

FACE-TO-FACE   TELEPHONE   ORTHOGRAPHIC   TRUE   SAMPLE
                                  DRAWING   PERSPECTIVE

Figure 9.1. Choose the Most Effective Method of Communication.

1. Verbal, face-to-face
2. Verbal by telephone
3. Verbal through third party
4. Written
5. Free-hand sketch
6. Orthographic drawings
7. Axiomatic, cabinet or cavalier projection drawings
8. Perspective drawings
9. Photographs
10. Samples of parts
11. Movies or video-tape recordings
12. Full-scale prototypes
13. Combinations of these

GUIDELINE NUMBER 4: PLAN THE TIMING FOR HIGH EFFECTIVENESS

A design engineer sent a letter to a production engineer to advise him that new equipment would be needed to manufacture a new product. When production was ready, it was discovered that the equipment had not been ordered. "But I sent you a letter a year ago," said the exasperated engineer. "Maybe you did," replied the production engineer, "and if you did, I probably set it aside because it only takes three months to get new equipment."

Here was the case of a conscientious design engineer, doing what he should be doing—conscientious, but unaware of the importance of timing. Too soon can be as ineffective as too late.

There is an optimum time for effectiveness on most design communications. You can become aware of it by putting yourself in the place of the recipient. Ask yourself, "When would be the best time for me to get the communication? When do I really need to know?"

GUIDELINE NUMBER 5: USE REDUNDANCY
IN THE METHODS OF COMMUNICATION

If there is one chance in ten that a communication can be misunderstood, then two communications of the same message by different methods have only one chance in a hundred of being misunderstood. This is the principle of *redundancy of method.* This is why we often follow up a telephone conversation with a letter, or vice versa. That is why the designer often pins down the design details with a specification, a drawing *and* a sample.

Be sure to use redundancy of method when communicating a new design technique. Remember that the onus is on you if there is a misunderstanding. With something new, the safest assumption is that your recipients are in a state of technical ignorance about what you intend by the design communication—to assume otherwise could be disastrous.

In my experience, I have known of many design problems which were simply due to a miscommunication. To those who are familiar with digital communications, I refer you to the parity bit concept where a confirmation code is transmitted immediately after a character stream. Those with a background in telecommunications will know that it is not unusual to use two separately routed telephone lines, or to have two different short-wave radio stations, to transmit the same data with very low error rates for the combination of channels.

GUIDELINE NUMBER 6: NEW THINGS
SHOULD BE PHASED IN GRADUALLY

We all know about resistance to change. It is always there to some degree. Yet, *the name of the game in design is to bring about change.*

A new state of affairs is created in every case, or if it is not, we are not designing—we are building more of the same. New designs range in a spectrum from minor changes like the special curve in a highway, to major innovations like computer controls on a manufacturing process. Let us consider what should be done in introducing a major innovation, then you can apply this rule to the degree that is merited by lesser changes.

In years of studying, reading about, writing about and implementing change, I have found only one universal rule: *Do it gradually!* The reason for this is that there are limits to what human beings can adapt to in the rate of change of their environment.

There is always resistance to change. It only becomes a serious problem when there is a major change in technology which affects the livelihood of those involved. We don't like things to happen too fast. In a major innovation, the information flow rate may be too high for us to accommodate. Our human defense against overcommunication is to resist the change, or to be confused and respond by copping out. Nevertheless, on the positive side, a small amount of change is both stimulating and interesting to us. This is the reason why innovations can, in fact, be introduced. It is fortunate that change is desired as well as resisted.

From a practical point of view, what you should do is to escalate innovation in small steps. It is surprising how much change can be achieved by this technique. The rate of change should always be at the point where interest is maintained but below the point where resistance is encountered. A small amount of resistance is permissible provided the results are beneficial. But whenever the resistance becomes too great, back off, listen, and replan.

Here is an example of how you should introduce a new product design which involves a big change in technology.

1. Announce the new product before the design starts—and beat the grapevine. Make sure that the advantages are understood.
2. About a month later, list the new skills required.
3. In the following week, announce the retraining and relocation plan.
4. Hold discussions involving the supervisors.
5. Revise the plans as indicated by the discussions.

6. In about the third month, provide lectures for the management executives about the general effects of the new technology on the organization.
7. Order new machinery for the facilities.
8. Train the workers in the new technology.
9. In the sixth month, have a trial using the new innovation in the production system.
10. Review the plan in light of the trial period.
11. Full operation in the eighth month.

GUIDELINE NUMBER 7:
INCLUDE FEEDBACK LOOPS

As previously mentioned, communication of the design is a process and not a step function. Any process is better if it has feedback to correct the output. A good plan is to have as many loops of feedback as you can in the overall design process.

A designer should encourage voluntary feedback by listening attentively and without overt judgment during all phases in the design process. He must take special care to not be defensive about the design or he will not learn what is happening in other parts of the organization.

Another way is to incorporate formal feedback loops: for example, by sending out letters with specific questions to selected people. You can get formal reactions from users, production people, salesmen, customers and others to ensure that their understanding of the new design solution is as it should be.

Much effective feedback is done by face-to-face communications about the design solutions while it is still under way. This should not be left to chance and you will be wise to deliberately plan to dig up predictive information. For example, you can examine the drawings, prototypes and the early production units and talk about the results with the people involved. You can make periodic visits to the field to question users and maintenance men. True, feedback loops may cause you to make minor design iterations—but it is to your advantage to do these as soon as possible instead of waiting for the roof to cave in.

APPLICATION TO AN EQUIPMENT DESIGN

In the case of the substation power transformer, which we have carried through the previous chapter, there is generally no communication problem of the design intent, because the design labs and the workshop are usually in the same building. Moreover, much of the success of the communication depends on past communication and accumulated know-how. In other words, the shop foreman is able to work effectively from drawings and correctly make the many interpretations of the drawings which are necessary.

With some new equipment designs, this is not the case. A design is completed and a set of drawings and specifications are sent to a contractor. Then we have problems. In the case of the power transformer of this example, I have chosen a phase transfer in the design process. The transformer has been completely manufactured and tested and is to be installed and maintained by personnel in another country. Just to increase the difficulty of communication, I will have the designers work in English and the installers work in Spanish.

A Communication Plan for Transfer to the Installation and Maintenance Phase.

1. Understandable to the Recipient.

- *Instructions must be duplicated in Spanish. Because of the argot and technical jargon, this is no ordinary translation job. The translator can be of a commercial variety, but all translation must be approved by a bilingual engineer who is familair with transformer technology.*
- *The English instructions, being the original ones, must rule in case of dispute.*
- *All drawings, installation instructions, and maintenance manuals are to be written in both languages in the same document.*
- *All labels on the equipment itself are to be in both languages.*

2. Arrange in Order of Use.

- *Get written descriptions from the installation and maintenance personnel on how they do their job. Use this to arrange the table of contents of the corresponding manuals.*

- *Have a duplicate set of unpacking instructions securely attached to the shipment itself.*
- *Redraw the assembly drawing for the manuals so that the hidden lines remaining will only be those with meaning to the maintenance person.*
- *Specify the special tools that are needed and locate the list in the front of the maintenance manual.*

3. Choose the Most Effective Method of Design Communication.

- *Face-to-face or telephone conversations about the design between English and Spanish speaking persons should be followed up with memoranda of understanding for each person's file and sent to a translator for comparison.*
- *Augment service manuals with plenty of annotated photographs.*
- *Augment orthographic drawing in maintenance manuals with exploded 3-D drawings.*

4. Plan the Timing.

- *Find out the address of the installation and maintenance contact three months prior to delivery.*
- *Mail preliminary manuals by air at the time delivery is started by sea.*
- *Revise and issue final manuals three months after first installation.*

5. Use Redundancy of Communication Method.

- *Have a maintenance electrician participate in the testing of the substation transformer in the factory prior to shipment. This person should preferably be the foreman of the installation and maintenance crew.*
- *Have a design engineer inspect the first installation when it is completed.*

6. Introduce Innovation Gradually.

- *There is no need to introduce this design gradually because it is a well-known technology.*

7. Plan the Feedback Loops.

- *The foreman who comes to the factory to witness the testing is to brief the design manager on the plans he has for installation and maintenance.*
- *Have a review meeting with the installation foreman and the inspecting engineer after the first installation. Their memorandum will go to the design manager, manufacturing and shipping.*

APPLICATION TO A CONSUMER PRODUCT

A Plan for Effective Communication for an Electric Toothbrush from the Design Phase to the Production Phase.

A small consumer products company will presumably have the advantage of short lines of communication within the organization. However, so as not to leave communication of the design intent to chance, let us look at the seven Guidelines for Design Communication which would most certainly be desirable in a large organization.

1. Understandable to the Recipient.

- *Esthetic standards to be communicated by samples.*
- *Ask foreman to mark up drawings in plain language.*
- *Make exploded assembly and service drawings.*
- *Obtain a dictionary of dental terms.*

2. Arrange in Order of Use.

- *Have tooling contractor rearrange the dimensions on the tool drawings before approval.*
- *Quality control specifications to be arranged in order of testing.*

3. Choose the Most Effective Methods of Design Communication.

- *Face-to-face meetings between the industrial designer and your contract customers.*
- *Photos of mock-ups for early product advertisements.*
- *Telephone to members of the dental association for their comments on the pilot run tests.*

4. Plan the Timing.

- *Make a list of main events and then schedule the timings of drawing issues, order placing, manpower changes, and advertising.*
- *Find out when UL can do their approval tests.*

5. Use Redundancy of Communication Methods.

- *Samples to go with drawings for construction of injection molds.*
- *Color samples to accompany specified paint codes.*
- *Specify bristle stiffness by bend tests and by limit samples.*

6. Introduce Innovations Gradually.

- *The product is simple, but the electrical and mechanical technologies may be new to the company. If so, organize an in-house training course for the production testers and inspectors.*

7. Planned Feedback Loops.

- *Make a follow-up schedule for major tooling.*
- *Make a pilot run of 1000 units and do a minor market test before full-scale production starts.*
- *Put customer questionnaire forms with the first 10,000 production units.*

APPLICATION TO A LARGE-SCALE CAPITAL PROJECT

A Hydroelectric Dam for Latin America

Assume that we have completed the concept phase. We are now ready to do a feasibility phase in order to obtain and analyze enough data so that we can let a contract for a detailed design phase.

Documentation:

Include a set of objectives, tentative criteria, and suggested general trade-offs between performance, cost and time. All correspondence and sketches are to be made available to the team doing the feasibility phase.

Other Media:

We will improve communications by having at least one member of the concept phase team posted to the next phase. A representative of the users in Latin America will also be asked to join the team.

Timing:

Although we are not yet ready to issue design or construction contracts, we will keep prospective tenders informed of our design decisions. The users will be asked to start informing the populations to be affected by floods. Get advance lists of trade skills that will be required for the project. A tentative schedule for all phases can be set up now and up-dated monthly.

Introducing Innovation:

Although irrigation is not a new technology, many irrigation projects in developing countries have had slow adoption of the innovation. The government of the users' country should be made aware of our data on this matter.

Planned Feedback Loops:

Because of the size of this design project, we will appoint a full-time liaison for keeping the client executive informed of the progress on the project.

A CASE HISTORY OF THE MISCOMMUNICATION OF A DESIGN DETAIL

An odd-shaped metal part was designed as a bracket for a mass-produced machine. It was to be formed and punched from sheet metal. Fifty thousand of them were to be made.

For some technical reason, the burr on one of the punched holes was to be on the opposite side from all of the others. The production department asked that the burrs be all on the same side so as to reduce the number of punching operations. After considering the consequences, a new design engineer in charge of the machine containing the bracket agreed to the change and the production department went ahead and ordered the tooling. A month later, the new design engineer issued an Engineering Change Order to bring the drawing up-to-date. The key line was:

Zone (E,5) delete: see note 12 and delineation

Under "Reason," the change order read:

Removed note specifying burr side of hole to make drawing compatible with parts made by existing tooling.

The E,5 zone of the drawing is shown in Figure 9.2. What actually happened was that the draftsman deleted not only "See note 12" and its arrow of delineation, but deleted the small hole as well. This

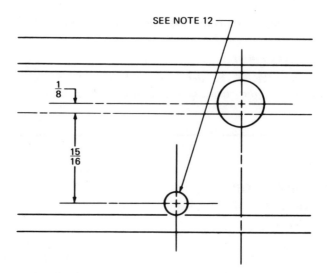

Figure 9.2. Miscommunication of a Design Detail.

change was also passed by the drawing supervisor who checked the drawing against the Engineering Change Order before it was issued.

Fortunately, this miscommunication had no serious consequences. The vendor who was making the parts noticed a discrepancy between the drawing in question and a separate subassembly drawing which showed the hole. The error was caught in time and corrected.

Comment: There isn't an experienced engineer in the world who has not run into some problems in translating ideas into real objects through the intermediate step of drawings. Even without errors, there can be misunderstandings. In this case history, the word "delineation" meant different things to different people. The engineer thought it meant the arrow, the draftsman, his supervisor, thought it meant the hole and its dimensioning as well. This misunderstanding was possible despite the parallel paths of communication provided by the "Reason."

What more could have been done? Obviously, a copy of the old drawing with some red pencil marks would have made the communication clearer by using a visual medium for a visual item. The drawing change could have been rechecked by the originating engineer. This is often done because people realize the difficulty in communication of engineering thoughts by words alone.

How far should one go in using redundant paths of communication in a situation like this? It depends on the consequences of an error. In this case, the cost of repunching and replating 50,000 parts would cost more than the annual salary of an engineer. Yet, if you are short of engineering staff, you cannot double-check every trivial change. It's a tradeoff. Because of the high cost of perfect communication, some errors in communication can be tolerated. The most effective check is *feedback,* either by drawings, prototypes or production samples. Above all, if members of the organization are prepared to accept that engineering communications are difficult, they will look for discrepancies and ask for clarification. This isn't fault-finding, it is good sound practice.

WRAPPING IT UP

When you spend months or years completing a design, you want to be sure it is carried out the way you intended it to be. Turning it over to other persons is not a step function. It is impossible to communicate that much information in a short time. It is a process you start before the handover and that you will still be involved in even after the sign-off date. The most important thing for you to realize is that the communication of the design intent is no easy matter. When you are aware of this, you can take sensible steps as shown in the applications. You will know that it must be done because a problem saved is time saved—even though you cannot prove it when the problem is not there.

Some of these guidelines will be new to you. To capture their essence, you should apply them to a design problem as soon as you can. After you handle the review questions, go to a project and make a plan for the design communication to another phase. This will give you a mental model that will help you get started when you face this in a real life situation.

DESIGN GUIDELINES

GUIDELINES FOR ENGINEERING DESIGN COMMUNICATIONS.

Do	Don't
Do make the information understandable to the recipient. The success of a design communication is the responsibility of the sender.	*Don't jargonize your enlightening hieroglyphics, or the party of the second part may not understand the proclamation of the first part.*
Do arrange the information in the order required by the user.	*Don't call the plays out of turn or you won't have a game.*
Do choose the most effective method of communication for any difficult items.	*Don't go after an elephant with a BB gun, unless you want it around your neck.*
Do time your communication so it gets attention.	*Don't excuse yourself with "better late than never" when never is better than later.*
Do use redundancy in the method of communication.	*Don't describe a beautiful sunset when you are at a loss for words.*
Do phase in gradually any radically new technology.	*Don't plunge into icy water when you could go in an inch at a time.*
Do include feedback loops as a check on the effectiveness of your communication.	*Don't depend on all your messages being "read and understood, sir."*

MAKING THIS KNOWLEDGE YOUR OWN

Questions About the Content of This Chapter

1. Who is responsible for the success of the communication between the design engineer and the people who make the design a reality?

2. Name two cases where arranging the communication in the order of use would be beneficial.
3. It's easy to understand how a communication can be too late. What is the rationale for saying that it can also be too soon?
4. What is the theoretical basis for using the Redundancy of Method for more accurate communication?
5. Why should an innovation be phased in gradually?

Questions on the Case History

A detail of a design, such as a hole, may appear on more than one drawing. Name one advantage of this fact and name one disadvantage.

Your Assignments for Becoming
a Top-notch Designer

1. For your practice project, make up a design communication plan for transferring it from the concept phase to a preliminary design phase (if this is appropriate). If you are working on a real project, make up a plan for transfer to the next phase, whatever it is.
2. If you or members of your group have a background in engineering design, table a few case histories for examination of how they were transferred to another phase. Report on how the application of the Guidelines may have improved the transfer.

10 Optimization is the Way to Get the Best

KEY IDEA: MODELS HELP TO FIND THE BEST VALUES FOR KEY
DESIGN PARAMETERS.

THE BEST IS YET TO COME

"Our units are ten times as effective as they used to be because we used optimization techniques," was a statement of an entrepreneuring friend of mine who managed a small firm which made high technology instruments. For years they had adjusted their design parameters by "gut feeling" and improved their products little by little.

Not much was known by anyone about producing sonic waves in water by thumping the airside of a diaphragm (called a "boomer"). Thumping the diaphragm harder was not the answer because of secondary pulses produced by undamped vibrations of the device. If a strong sharp pulse was to be produced, more energy was not the answer. The usual assault on this problem was the brute force technique of trying a variety of thumping mechanisms. They had only reached an efficiency of .01 percent by such techniques. However, the manager of this firm could see no reason why the efficiency needed to remain at such a low value. He hired a mathematician and physicist to develop the equations of the device as a system. After one year on the first effort and $60,000 worth of computing time at $300 per run, they called it quits and went back to the starting point. They hired a new mathematician to take a different approach to developing equations that could be solved by heuristic methods on a computer. They also contacted a university research unit and found that one of the university professors had just developed a new heuristic for solving optimization equations. The university professor was contracted to assist the mathematician and within a year they had some outstanding results. The computing time cost dropped to $30 per run. (It took hundreds of runs to explore the key parameters.)

What they were able to show through their optimization analysis was that the theoretical efficiency of their unit could be very much higher than the current units . Within a few months they had built a new prototype with more optimal type of dimensions and had achieved the ten-fold increase that was mentioned at the beginning of this chapter.

Another result of their optimization analysis was that they were able to persuade their major customer, the government, to award them substantial sums of money for the development of better units. Their customer could see that their optimization technique was the reason for their optimism. They• were not just fooling around.

WHAT YOU WILL GET OUT OF THIS CHAPTER

Optimization is a process that can be applied in several design phases, or as a parallel effort. It is a search for the best parameter values in the early phases, and an overall optimization in the final phase. But it only happens if you want it that way. It takes extra effort to get that extra plus.

Objectives

1. *You should learn that there are two kinds of optimization.*
2. *You should learn that there are many kinds of models to aid in optimization.*
3. *You should learn about computer-aided optimization techniques.*
4. *You should learn the steps to take for optimization.*

Benefits

1. *When called on to produce the very best, you will know how to go about it.*
2. *You will be able to decide when a computer can help with optimization.*
3. *You will be able to turn a decision matrix into an optimization tool for going from the mediocre to the excellent.*

At the beginning of this chapter, you saw how an optimization effort paid off with a fantastic improvement in performance. Now I will show you how it is done and when you should do it. Five ways are discussed, all the way from "cut and try" to linear programming with a computer. Suggestions are made of ways to go about optimizing an equipment design, a consumer product, and a large-scale capital project. The two case histories show the range of optimization from furniture to an echo-sounding device. At the end of the chapter are questions and assignments to guide you into the practice of optimization.

THE TWO MAIN CONCEPTS OF OPTIMIZATION

The True Optimum

You may wish to find the maximum or minimum of some single design criterion, knowing that such a condition exists. For example, you may wish to have maximum performance, or maximum efficiency, or maximum yield of a process, etc. On the other hand, you may seek a minimum of heat loss, distortion, or surface area of given volume and so forth. This is the kind of optimization shown in Figure 10.1. There is a theoretical maximum (or minimum) and you are out to achieve it.

An Optimal Solution

You may wish to maximize (or minimize) some criterion as far as you can within the time, technology, and resources available. There may or may not be an absolute maximum or minimum. Examples of this concept are the maximum esthetic appeal, the maximum

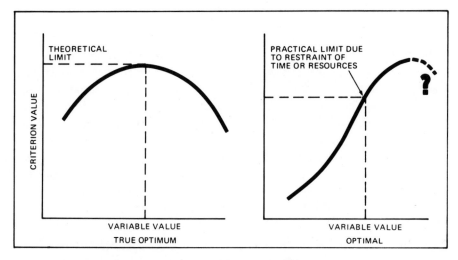

Figure 10.1. Two Kinds of Optimization.

profit, the minimum cost by value engineering, minimum maintenance cost, minimum wear, and so forth. You can reach an optimal
design with respect to some "ideal" objective which can be reached
with a reasonable investment of time and resources. This concept
is illustrated in Figure 10.1.

The Principle of Design Optimization.

*The best design may be approached by careful selection and/or balanc-
ing of the design parameters by achieving the best of a criterion function.*

WHY SHOULD YOU OPTIMIZE?

An amateur designer is satisfied with anything that works. The professional designer not only wants to get the best under the possible
circumstances but knows how to do it. Therefore, one good reason
for optimization is the simple self-satisfaction that comes to the
professional from doing the best that can be done.

Ego points alone are not good enough reasons for optimization. There should be benefits to the users, designers and all those involved which warrant the extra cost in time and resources which are generally required to get an optimized result. In other words, what is better must be perceived by the users as having some tangible value. In consumer products, a better product commands a higher price and therefore pays the design team in terms of their share of the profits. Optimization of a consumer product gives you the little extra edge which makes you competitive, be it the best for performance or best for cost. On the other hand, it is difficult to justify an optimization effort on a standardized girder of a bridge, or a standard curve on a highway cloverleaf. The optimal solutions for these are known and the experienced engineer knows what to do.

By far the most difficult optimization to justify is that of a socio-economic technical system such as a hydroelectric dam and irrigation system. The overall optimization of combined social and material benefits is as yet an infant science, although it has been done in a number of cases.

It boils down to this. Optimize as much as can be justified by the increased value obtained from the results.

WHEN SHOULD YOU OPTIMIZE?

Optimization Before Selection

After analyzing for feasibility, you may not be certain which is the best solution. You may need to do some optimization with respect to the important criteria so that you can clearly discriminate between closely competing alternative solutions. This is one type of optimization and it is sometimes inserted as a formal step between the screening for feasibility and the selection of the solution.

Post-Concept Optimization

When you have selected a concept, so that the components of the system are known, it is possible to set up the relationships and study their interaction. Actually, in most cases optimization would occur at the preliminary design phase which follows the conceptual phase and precedes the detailed design phase.

Post-Design Optimization

After a design has been completed and working for a while its operation is better understood. A little of the fat is trimmed off the meat by reducing safety factors to more reasonable values. On a complete design it is possible to reduce the safety factor a little bit and check with a test. Such post-design optimization would result in both better performance and lower cost. For example, an automobile engine which has a history of five years of use and improvement is often said to be optimized.

Optimization as a Process

The concept of modeling and improving a design can be a continuous process at all phases in design. Suppose, for example, you are designing an off-road vehicle and comparing alternatives for traction. You may optimize each of the concepts by shifting the engine weight until it is clear which alternative would really be best. Having selected one major concept then in its detailed design phase you may optimize again with respect to traction by carefully selecting the parameters for wheel loading, width, diameter, torque, etc. And, finally, after years of experience you may introduce a new rubber tread with an optimal type solution for traction on pavement, mud or sand.

HOW DO YOU OPTIMIZE?

Cut and Try

Experience takes the risk out of guesswork, and many gadgets and unusual machines are best optimized by applying expert judgment on which changes should be tried out. Anyone who attempts to optimize a potato peeling machine by cut and try techniques had better be backed up with some expertise on the machine in question, because the models used for optimization are in the mind.

An Iconic Model

Any physical model which resembles the final design, be it scaled down or full sized, is an iconic model. They are excellent for optimizing the appearance, the convenience, the human factors of usage, the fitting together of unusual shapes and so forth. Scaled-down models are great for optimization tests on things that cost a lot to build such as aircraft, ships, new factories, shopping centers, etc.

A Graphic Model

Suppose you are designing a factory and you need to provide room in the aisles for moving in and out a large, unusual shape of machine. A scaled-down graphic drawing is about the only way to find out the minimum size of aisles and corners that will be required. Likewise, an optimal floor layout can be determined by comparing graphic models of alternatives.

A Mathematical Model

Calculus is the tool for finding maximums or minimums when you are able to construct a mathematical model or part of, or all of your design. In a practical sense only part of the performance of a design can be mathematically modeled, but there are great possibilities in this technique. Dynamic mathematical models have been highly developed in electronic science and have been applied to mechanical systems as well. Using these techniques, models have been developed for chemical processing systems, and extended to some extent to economic systems. The science of mathematical modeling has been extended to social systems, such as the population model developed by J. W. Forrester, *World Dynamics,* Wright-Allen, Cambridge, Mass., 1973.

Practically speaking, a true maximum or minimum can only be found between two related variables. Since almost every design has more than two variables involved, then we have a multidimensional surface to explore for peaks and valleys. There is some ambiguity as to which is the "best" solution under these circumstances. A lot of progress has been made with hill-climbing techniques which enable you to find a high hill even if you do not find the highest. In other words, even with multivariable optimization, one can find better solutions without necessarily finding the ultimate solution.

Linear Programming

When a system can be modeled by a set of linear equations with restraints on most of the variables, an optimal type solution can be found using computer programs which are universally available. The general types of design problems which lend themselves to linear programming are similar to the following examples:

1. Alternative combinations of mixing chemicals, cattle feed, food, colors, etc., may be analyzed to find out which combination gives the lowest cost.
2. The capacity allocation of a factory to various products which are optional can be analyzed and the optimum combination found for the maximum profit. If good records are kept, a factory can determine which product mix will be most profitable for them. In the practical sense, suggestions can be made to the sales department as to the most profitable combination of sales.
3. Production scheduling is another use of linear programming. The juggling of sales demands, inventory and storage costs can be done to maximize the profits.
4. The routing sequence for trucks, ships, and other means of transport can be optimized to lay out the most economic routes and time schedules.

Linear programming finds its greatest utility in operations and processes which are linear, and where good historical data is available. There are also nonlinear programming techniques available and these have more potential for use in engineering design.

AVOID THE PITFALL OF SUBOPTIMIZATION

The subsystems of a system under design can be optimized separately. If this is not controlled it may upset the balance and reduce the optimization of the system as a whole. For instance, the transmission of a car is a subsystem which can be optimized with respect to the criteria for the whole car as a system. The speed and torque conversion are optimized with respect to the overall system criteria for weight, size, etc. If one optimizes the transmission solely for its own sake, say to have a 16-speed transmission for a passenger car, this is suboptimization which is counter-productive to the optimization of the whole car. Suboptimization sometimes occurs because some part of the design team believes that their component is more important than the other components.

To avoid this pitfall, optimization should be done with respect to the total design objectives.

APPLICATION TO AN EQUIPMENT DESIGN

Many equipment designs are one-of-a-kind, where performance is to be optimized and not the cost. For these it may be appropriate to use scaled-down models and graphic models to improve the expected operation as much as is possible with the time available. The result could then be called the optimal solution.

EXAMPLE OF AN OPTIMIZATION OF
A CONSUMER PRODUCT—
THE ELECTRIC TOOTHBRUSH

The total satisfaction obtained from a design can be expressed as a weighted sum of satisfactions obtained from meeting various design objectives. This was covered in the Decision Matrix shown in Figure 10.2. Here we see that a design alternative resulted in 71 percent satisfaction of overall objectives. Now, by noting the lower triangles of each box we see numbers which correspond to the contribution towards each objective. This number is equal to w times X. It depends on the weight and percent satisfaction of the objective.

What would it take to add more to the number in the triangle and raise the 71 percent Overall Satisfaction? A sensitivity analysis of criterion function, shows that the potential for increase in satisfaction

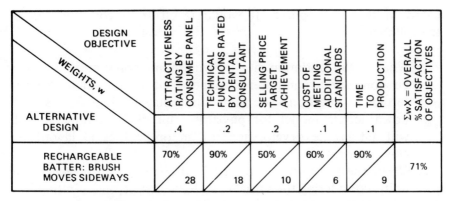

Figure 10.2. Example of Optimization of a Consumer Product.

will come from those having high weight and a low percentage satisfaction. For example, the design objective "technical functions rated by dental consultant" has a weight of .2 and a percentage satisfaction of 90 percent. Although the weight is high, there is not much room for improvement here. On the other hand, the objective "selling price target achievement" with a weight of .2 and percentage satisfaction of 50 percent shows considerable room for improvement. This suggests that the optimization should be directed toward reducing the selling price for the same level of technical functions. This could probably be done at a small loss in the satisfaction to the objective "time to production" which only has a weight of .1. In other words, it looks as if a little effort on cost reduction by the technique of value analysis is warranted.

APPLICATION TO A LARGE-SCALE CAPITAL PROJECT

A Hydroelectric Dam for Latin America: Optimization

As this example is being done as a concept phase, no specific optimization is possible. However, an optimization plan can be prepared in this early phase. It would result from consideration of the following:

1. What is the optimum site for minimum capital cost? This can be done by a computer simulation of various dams for the terrain in question. The same study may be done by a scaled-down iconic model for physical simulation. Graphic models will aid in the development of an optimal canal layout for gravity irrigation.
2. Given the site, what is the optimal combination of generating units, considering the initial cost, breakdowns, and additions?
3. Given the site, and the dam height, what is the optimum combination of local materials and building techniques for a minimum construction cost?
4. If steel pipes are used to conduct water from the dam to the power house, what is the optimum diameter for minimum overall cost?
5. There is a major tradeoff in water for upstream irrigation and the generation of power. Data will be required on the technical tradeoff, as well as on the value functions of the users for irrigation and for power.
6. The overall optimization will be to seek an optimal solution, but this may include an optimum of some subsystem or of some major variable.

CASE HISTORY OF AN ICONIC MODEL
USED FOR JUDGMENTAL OPTIMIZATION
OF A CONSUMER PRODUCT

An industrial design organization had to make a tradeoff between the comfort and stacking attributes of their ottomans (seat and foot rest). These consisted of an inverted plastic box with sloping sides and topped by a cushion. Placed side by side they would substitute for a couch and when stored they could be stacked one on top of the other.

The steeper the sides, the more comfortable in the couch arrange-

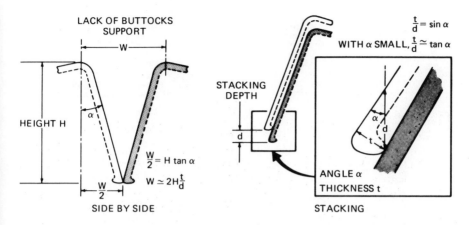

Figure 10.3. Ottoman Optimization.

ment but the more storage space required for stacking. Although the situation lends itself to the mathematical treatment as shown on the sketches of Figure 10.3, the solution lies entirely in the area of judgment. What the designers did was to make a full-scale mockup (iconic model) of two of them and try them for stacking and for seating comfort. Since time and resources were limiting factors, a compromise solution was arrived at. We could call it their "optimal" solution.

A CASE HISTORY OF A COMPUTER HELPING A "BOOMER"

At the beginning of this chapter, you have read about the results of using a computer to solve the mathematical model of a sonic boomer. How was this done? Here are some of the details, abstracted from a paper presented at:

Ocean 74, IEEE International Conference on Engineering in the Ocean Environment, Halifax, Nova Scotia, August 21–23, 1974. Figure 10.4 is a diagram of the apparatus and the corresponding equation of motion.

When the initial conditions were applied, eight state equations were developed. To solve these with computational efficiency, a

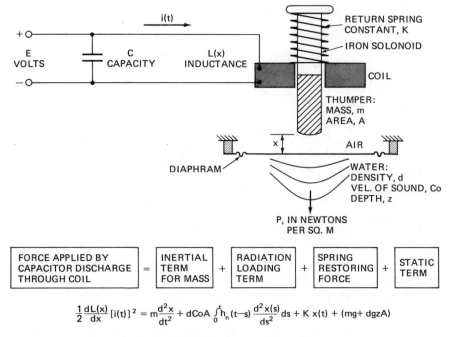

$$\frac{1}{2}\frac{dL(x)}{dx}\left[i(t)\right]^2 = m\frac{d^2x}{dt^2} + dCoA \int_0^t h_n(t-s)\frac{d^2x(s)}{ds^2}ds + K\,x(t) + (mg + dgzA)$$

Figure 10.4. The Boomer Equation.

heuristic solution was needed. At a time when they were faced with very high computing costs, they were fortunate in finding a university professor who had developed a new algorithm for an optimization computer program. It was exactly what they needed.

As usual, with an optimization model, this one was validated by experimental data taken from their current production model. The results are shown in Figure 10.5.

OPTIMIZING YOUR GAIN FROM THIS CHAPTER

Think about optimizing your engineered designs whenever this practicable. Excellence in engineering is the plus that comes from the extra effort. Now you know about the tools to guide that effort. Follow through by doing the appropriate exercises at the end of this chapter. Then optimize a real project and get a winner!

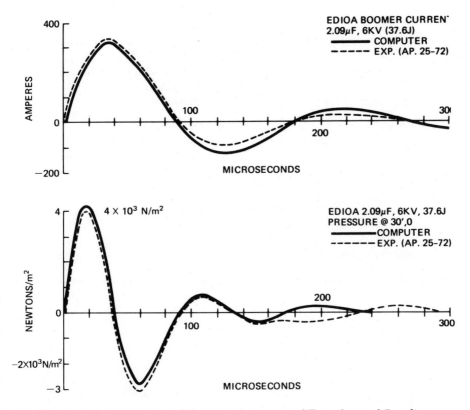

Figure 10.5. Comparison of Computed and Actual Experimental Results.

DESIGN GUIDELINES

GUIDELINES FOR DESIGN OPTIMIZATION.

Do	Don't
Do optimize the overall design with respect to the design objectives.	*Don't overoptimize the part you like to do. You might suboptimize the whole.*

Do develop a criterion function by (1) finding the mathematical relationship or (2) identifying a dominant parameter or (3) a linear criterion function of weighted satisfactions.

Don't back off because of the effort of developing a mathematical model for (1). More and more people are finding it worth the effort. Anyway, you can always use (3).

Do use your skills to find the optimum or optimal value of your criterion function.

Don't rely on mathematics. Nothing was ever designed by sophisticated equations. You must reduce them and solve them so that it helps the design effort.

Do choose those kinds of models that will enable you to make predictions and judge the effect of parameter changes.

Don't try to do it all in your head. You could blow a fuse.

FOLLOW-UP AND PRACTICE

Questions on the Content of the Book

1. What is the difference between an optimum and an optimal type solution?
2. How many kinds of optimization models can you recall? Name them.
3. Name at least two optimization techniques.
4. How is a Decision Matrix used for guiding the overall optimization of a design?

Questions on the Case Histories

1. At the beginning of the chapter, there is part of the case history of the echo-sounder. More details are near the end of the chapter under the title "Case history of a computer helping the boomer." What was the reason that they had to abandon the first effort at optimization? What were the key factors leading to the final success?
2. How was the mathematical model validated?

Assignments for Becoming
a Pro at Optimization

1. For your practice project, or your real project, make an optimization plan that contains every applicable optimization technique. Assume that generous amounts of time and resources would be available for the optimization effort.
2. To expand your mind to other optimization techniques, examine some past engineered designs that you are familiar with, and make up a proposal for a redesign effort using applicable optimization techniques.
3. When you do get a winner through optimization, write an article or give a paper on it. The engineering world needs to get with it on optimization techniques.

11 How to Control the Time and the Cost

KEY IDEA: TO BE IN CONTROL, YOU MUST HAVE A DETAILED PLAN OF THE ACTIVITIES, THEIR SEQUENCES, AND THEIR ESTIMATED COST. THEN YOU MONITOR AND REVIEW THE PROGRAM. YOUR EMPHASIS IS PUT ON THOSE WHICH ARE CRITICAL FOR TIMING.

MAKING IT WORK IS NOT ENOUGH

The engineering manager of a team of forty persons was responsible for the design of custom equipment. If the designs were not completed on time, or if the design or manufacturing cost was in excess of estimates, it was he who was held to account.

Speaking strictly from his point of view, he felt that not enough engineering designers were concerned about the time and cost aspects of a design. They were so concerned with technical perfection that they never would finish on time if it were not for management's actions. As he said, and other managers have said, "The design engineers never consider that they are finished, nor do they consider anything less than technical perfection to be important."

Well, that is, nearly everyone. The complaint is more often directed at the younger persons who are entering the practice of engineering design. They are hired because they are smart and it seems to be up to the employer to get them to realize that the cost and the timing of a project are just as important as how it works. It is not enough to make it work, it must be ready on time and it must be reasonable in price. The pros know this, that is why they are still in business.

Fortunately, for those who agree with the point of view of management about the time and cost aspects of engineering design, there are tools and techniques that make this possible. A whole new methodology has been developed for this purpose. It is known as *Project Management.*

WHAT THIS CHAPTER WILL DO FOR YOU

Objectives

1. *You will learn how to plan a PERT activity logic diagram and develop a time schedule for a design project.*
2. *You will learn how to estimate by activities in order to make a project budget.*
3. *You will learn how to control time and cost by milestone reviews.*

Benefits

1. *You will know when and how to initiate plans for time and cost control.*
2. *You will have a useful tool for scheduling design projects.*
3. *You will be able to keep a project on target.*

In this chapter, you will learn about the science of work planning and control. It is called PERT, which means Program Evaluation and Review Technique. A similar technique is CPM which means Critical Path Method. You will be shown the PERT/CPM technique. With it, you can set the stage for meeting the time, cost and performance targets for your design project. A step-by-step procedure is explained.

First you are shown how the engineering design process and project management should dovetail. Many people fail to realize how important this is.

Next, you learn how to make a logic diagram so that the project activities will be done in the correct sequence. You learn how to make a realistic schedule which you can use for getting the work done on time. Schedules need control actions if the project is expected to be completed according to schedule. Therefore, you must learn to make use of your knowledge of the engineering design process to have review meetings at the proper milestones.

The applications of Project Management techniques are built into the realistic example used to describe the technique.

There are review questions and exercises at the end of the chapter which will help you to become proficient in time and cost control.

DETERMINE THE PHASES AND MAJOR MILESTONES

When starting a project plan, you should first determine where you are in the engineering design process. If you are about to implement a completed design, then you have only one major phase to complete. You know what needs to be done to complete the job and developing a work plan will be straightforward. If, however, you are upstream in an early part of the engineering design process, you must ensure that you have a plan of formal iterations which will reduce the risk of a downstream disaster (See chapter one).

First, you should lay out an overall masterplan as shown in Figure 11.1. Suppose you are at the beginning of the "Prototype Phase" of the design of a compact stereo. You could plan out this phase in some detail but only rough in the later phases because they depend on the outcome of the prototype phase. In fact, your plan must acknowledge that at a decision review point, there is always consideration given to aborting the project right then and there— better earlier than later. You also consider repeating that phase, as we say, "back to the drawing board." If all has gone well in the prototype phase, then you go on to the next, having reviewed and revised the project objectives in the light of experience. A "decision review" is more than a design review. A major decision must be made.

Many projects for engineered designs fail because of a failure to plan the formal iterations as above. Don't get caught in that trap, no matter how much pressure there is on having early completion.

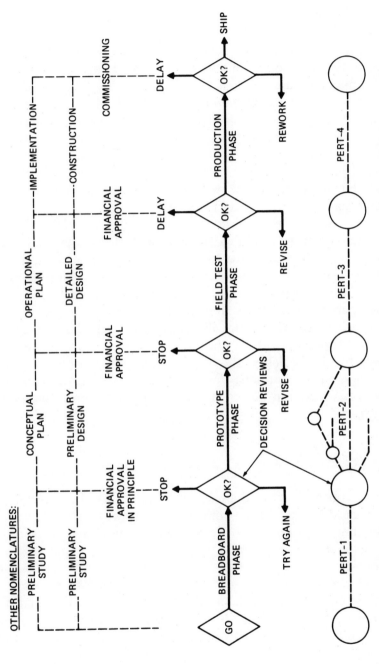

Figure 11.1. Phases and Decision Reviews in the Design Process, with Corresponding PERT/CPM Controls.

LIST ALL THE ACTIVITIES

An "activity" is a definable item of work on a project. How small the activity is will depend on how much control you need. You start with the major items, say, ten to twenty activities. You ultimately enlarge the activity list to hundreds or thousands as you proceed downstream. Right now consider yourself to be at or near the beginning. You want the engineering design process to work for you. Then be sure to include activities for all the important steps in the process. You start with the seven steps first listed in chapter one, consider what should be done for each, and list specific activities which are to be done (See Figure 11.2).

As you proceed from phase to phase in the engineering design process, the work gets defined in more and more detail. Large activities are broken down into subactivities, and these in turn, into small units until you get one-person tasks. This procedure is called the Work Breakdown Structure. Generally we proceed from system to subsystem to work type, as you can see in Figure 11.3. Tasks which are given out to individuals (or contractors) who have special skills, are at the bottom of the hierarchy. By encoding the tasks according to the speciality, they can be grouped together for assignment or combined contracts. For example, A1.4 and A2.4 are both mechanisms which can be assigned to the same team of mechanical specialists.

DRAW A LOGIC DIAGRAM

Whenever you have more than ten activities, it will be necessary to draw a logic diagram (the network) in order to work out the shortest possible time to complete a project—unless the activities are completely sequential. This is part of the PERT/CPM technique.

The question that must be answered for each activity is: *Which activities must immediately precede the one I am working on?* If activity C must be preceded by A and B, then this is all the information a computer needs to draw out a logic diagram for you. However, in the early planning stages you will do better to check out your precedence logic with a pencil diagram. Each activity is represented by an

Design Step	Kinds of Activities Needed	Specific Activities to be Included in a Prototype Phase for a Stereo
ANALYZE NEEDS	–A market survey with a mock-up –A human factors study –Demonstrate and get user comments	–Develop a product plan with target for performance, cost and time –Meet with marketing and customer
SET OBJECTIVES	–Meetings with funding authority –Review of past projects	–Set subobjectives for the subsystems –Get agreement on tradeoff guidelines –Get approval of objectives –Make project logic diagrams and baseline schedules
DEVELOP TARGET SPECIFICA-TIONS	–Write up draft specs –Conduct tests to determine constraints –Consult with experts to get parameter limits	–Draft system and subsystem specs –Get approval of specs from customer –Measure up competitive stereos
CREATE ALTERNA-TIVES	–Idea generating sessions –Patent searches –Literature reviews	–Build alternative design for amplifier subsystem –Order long lead time tooling
SCREEN FOR FEASIBILITY	–Economic analysis –Review meetings	–Test whole system on 5 prototypes
SELECT THE SOLUTION	–Decision meetings –Test evaluations	–Compare and choose between alternative amplifiers –Demonstrate to marketing
COMMUNI-CATE THE DESIGN SOLUTIONS	–Define final performance levels, materials, dimensions –Define quality levels by samples	–Confirm orders –Issue specs and drawings –Set future milestones –Issue PERT and schedule for next phase –Set aside a working prototype systems

Figure 11.2. Some Activities Come From the Engineering Design Process.

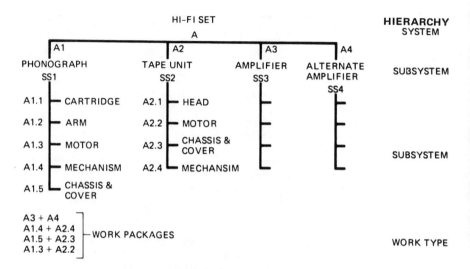

Figure 11.3. Example of an Hierarchical Work Breakdown.

arrow which has no duration in time but has a beginning and end point. The end point leads to a circle which represents the completion of the activities leading into it. To keep the precedence logic correct, a "dummy" activity is sometimes required. It takes zero time. A dashed arrow is used. Such a diagram is shown in Figure 11.4. It is a good idea to write the names of the activities above the arrows because errors in logic are difficult to perceive if you only number or letter the arrows.

When the logic diagram is completed to your satisfaction, insert the times below the arrows and see if you can determine the longest path through the network. (This will not be easy if your network has more than twenty-five activities. You may be required to use more sophisticated techniques, such as developing a "float table.") Note that the logic diagram is not drawn to a time scale. The critical path is the longest elapsed time path through this network, and therefore, is the shortest time in which the project could possibly be completed. The critical path is marked on Figure 11.4.

There is another kind of logic diagram which has been widely used on large construction projects. It is called Precedence Diagramming. The activity is a labeled rectangle. Within the rectangle is space for the duration time, its earliest start, latest start, responsi-

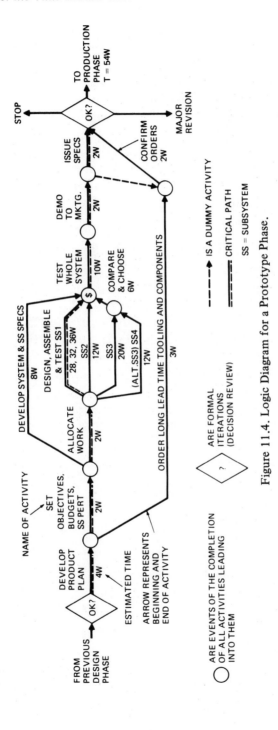

Figure 11.4. Logic Diagram for a Prototype Phase.

bility and other data. If a milestone is required, an activity rectangle is so labeled. All rectangles are joined by lines that show the logic. See Figure 11.5.

I have used both the arrow-node diagrams and the precedence-boxes and they both work. A computer can draw out the latter better than the first. The precedence diagramming is easier to learn for simple logic diagrams. However, when you get further into complex logic diagrams and "float" calculations, there is about as much to learn with one system as with the other. For this reason, the examples used in this book are arrow-node diagrams. Float will be dealt with later.

DEVELOP A BASELINE SCHEDULE AND ADJUST FOR THE AVAILABILITY OF RESOURCES

So far, we have assumed that the resources of people and equipment will be available whenever needed to complete the project in the shortest possible time. Sometimes this is true and the final schedule

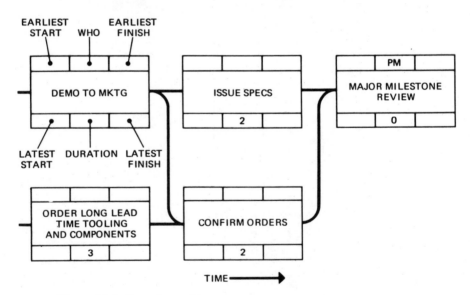

Figure 11.5. Precedence Diagramming: a logic diagram with boxes.

can be made up immediately. This is what we call a "baseline schedule" and it would look similar to that in Figure 11.6. This bar chart is also known as a Gantt chart.

A bar chart is a tool for *planning* and *control*. It gives a visual model of what is required and is more effective than a schedule of dates alone. On the other hand, if hundreds of activities are involved, a table of dates allows the information to be compressed into a smaller space than is required for a bar chart.

Note that the uncertainty in the time required for subsystem 1 has been indicated on the diagram because this is on the critical path and the activity duration is long. Uncertainty on noncritical activities is seldom as important. In the PERT method, we sometimes get estimates of the "optimistic" and "pessimistic" times as well as the "most likely."

The concept of optimistic and pessimistic times is most useful for upstream management, especially for R&D. You ask the engineer or scientist, "What is the most likely time to completion of this activity, in the normal course of events." Call this t_m. Then, "If things go well, what is the optimistic time?" (One chance in 1000.) Call this t_o. Finally, if things go badly, you will have one chance in 1000 to be as late as . . .? (Not forever!) This is the pessimistic time, t_p. Statistically, we can show that the single point time to use, called the equivalent time, t_e, is:

$$t_e = \frac{t_o + 4\,t_m + t_p}{6}$$

However, the reduction of three times to one leaves out the most important factor, the uncertainty. There are statistical ways to treat this matter, but they require expensive runs on a computer. A very practical way is to find out from your real time schedule where the uncertainty would matter, find out its magnitude, and record it on the bar chart. Then, do something about it. In Figure 11.6, SS1 is the one where it matters, and the astute project manager will do something about it, such as, a head start, milestones, additional resources or a contingency plan. That's management!

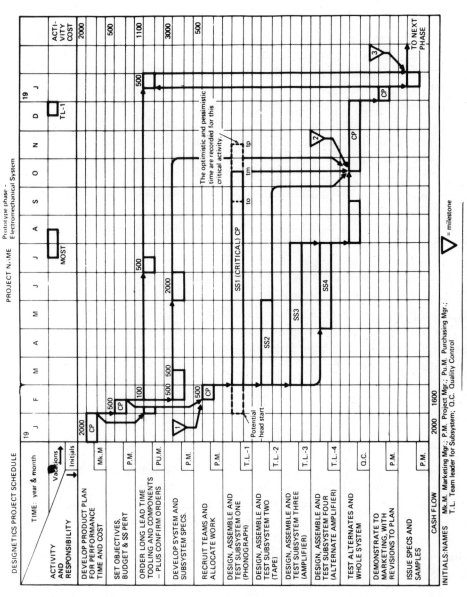

Figure 11.6. Project Schedule.

IDENTIFY THE CRITICAL ACTIVITIES
AND THE FLOAT

The critical path on the logic diagram of Figure 11.4 can be found by inspection. It is marked with a dash beside the arrow. The earliest finish of the total project is 54 weeks. (If we allow for the uncertainty in SS1, it is 54 ± 4 weeks.) For the critical activities, there is no extra time if the project is to finish as soon as it can—54 weeks. If there is spare time on an activity, it is called "slack" or "float." You can see that SS2 is waiting for SS1 to be completed and its float is 32 – 12 = 20 weeks, provided that it is started as soon as it can be.

Float is better determined when the real time schedule is ready. Thus in Figure 11.6, SS2 still has 20 weeks of float, which is "free float" because no rescheduling of the following activity is required. Look at SS4. Originally it had 20 – 12 = 8 weeks of float while waiting for SS3 to be completed, but because it starts four weeks later than the earliest start possible, there is now no "free float." Yet, because the "Test Alternates . . ." has four weeks of float, so do SS3 and SS4, provided that it can be rescheduled later. It is not "free." It is called "interfering float." The sum of the free and interfering float is called the "total float".

If you have many activities to coordinate and monitor, knowledge of free float and interfering float is very useful. If something is late, you know what the effects of rescheduling will be.

HOW IS CONTROL EXERCISED?

In order to get the total cost of a project, you should first estimate the cost of the individual activities. Some of the estimates will be quite accurate. This is the principle of estimating—break it into small parts and aggregate them.

In Figure 11.6, the estimated costs of the first five activities have been entered and summed by activities and by the project itself. The monthly cost is necessary for overall financial planning, which generally requires a projected cash flow.

At this point, you should review any milestone points that you tentatively set on your logic diagram and check on costs which are added up to give milestones which are significantly different from each other. Cost milestones are different from periodic milestones. If the schedule should change, so would the date of the milestone review of cost. Some milestones are indicated by numbered triangles in Figure 11.6.

Because of delays in some activities, a project cash flow may be underspent, but the money spent for the work done is still committed when the milestone is completed. Similarly, the early completion of an activity might make the monthly cost appear to be above normal, whereas you will be happy that something got done ahead of schedule for a change. (It does happen, you know!)

HOW IS CONTROL EXERCISED?

A degree of control is possible because all activities are scheduled for time and for cost. At a milestone, we can total up the money spent. We can take a look at the time elapsed, we can take a look at the performance we got for the effort, and we can do any necessary replanning to meet the basic objectives (targets) for the project. Milestone reviews and major phase reviews are somewhat similar, except that milestone reviews occur more often. They enable us to replan if we find the target moves around a bit. On the other hand, major phase reviews are not a replanning of the forward motion of the project but a review which includes consideration of whether the project should continue to go ahead or not.

In addition to milestone reviews, the updating of schedules for time and for cost enables people to perceive in advance that something is going extra well or is getting a bit bad.

Balanced control of time, cost and performance should be the outcome of any control procedure that you use on a project. Just as it is not enough that a design work, it is not enough that it be on time; and it is not enough that the cost be exactly right. It is the *balanced completed package* of performance, cost and time which is important.

The *tradeoff control* will determine the final performance, time and cost reached when upstream in a project. You can control each upstream activity itself by the methods shown in this chapter, but the effects of upstream activities on downstream results must be controlled by tradeoff guidelines. See chapter five for more details on how to do this.

MANAGEMENT OF A PROJECT

People with experience in project management who attend my seminar-workshops report one important fact: "It's getting people to do the work that is the problem—not the logic diagrams." I agree. The science of work planning and control, PERT/CPM, should not be mistaken for "management" of a project. It requires skill and tact in dealing with people, especially nonsubordinates. You will find some of this in the next chapter. It will help. The other factor is authority. You must have certain rights, or you will end up as a coordinator who is able to report on disasters, but not able to prevent them.

As I said, there is much more to project management than diagrams and schedules. Still, it's a good place to start. Happiness in design is having a project under your control.

STEPS FOR DEVELOPING A TIME PLAN.

1. *Determine the overall phases with decision reviews and other milestones that are desirable for completing the design process.*
2. *List all the activities required for completion of the next phase you are about to start.*
3. *Draw a logic diagram which shows which activities must precede others.*
4. *Develop a baseline schedule on the basis of all persons and equipment being available when needed.*
5. *Adjust schedule to availability of resources.*
6. *Identify the critical activities for which there is little or no float time.*

STEPS FOR DEVELOPING A COST PLAN.

1. *Estimate the costs for labor, overhead, and materials for each of the activities in the time plan.*
2. *Using the actual schedule, sum up the expenditures by months to get a cash flow forecast.*
3. *Sum up the cost of groups of activities preceding major milestones and the end of the project phase.*

DESIGN GUIDELINES

GUIDELINES FOR TIME AND COST CONTROL.

Do	Don't
Do make a detailed plan of activity, time and cost for each task in a project.	*Don't expect to have control by eyeballing the work except when it is routine.*
Do measure your progress against the plan by milestone reviews.	*Don't waste your effort on periodic reviews that are not at milestone points.*
Do be sure that the controls are balanced to achieve the correct package of performance, time and cost.	*Don't hammer too hard on the time aspects. PERT/CPM can overwhelm people with emphasis on time.*
Do control the process when upstream on a project.	*Don't try too hard to control the work to be done except when downstream in a project.*

Questions to Help You Consolidate Your Understanding of Time and Cost Control

1. Why should a project plan for an upstream project have design iterations?

2. What is the purpose of drawing a logic diagram before scheduling with a bar chart?
3. What factor can switch the critical path to other than that
   found with the logic diagram?
4. Go to Figure 11.6. How much float is available on the activity,
   ". . . Subsystem Two."

Assignments to Help you Gain Proficiency
in the Use of PERT/CPM

1. For your practice project, or real project, develop a time
   plan which includes a logic diagram and a bar chart schedule.
2. For SS3, which is an electronic amplifier, make a list of about
   a dozen probable activities and draw a logic diagram for it.
   If you are not familiar with this kind of item, try SS1, which
   is mainly mechanical. Find a colleague who can comment on
   your list of activities and their logic diagrams.

12 The Management of Engineering Design Teams: It's Different

KEY IDEA: THE MANAGEMENT OF DESIGN TEAMS REQUIRES AN UNUSUAL DEGREE OF FLEXIBILITY IN MANAGEMENT STYLE.

SHOULD THE MANAGER OF ENGINEERING BE A TECHNICAL PERSON?

When I was a young engineering supervisor for a large international corporation, I was lucky enough to be part of a group of twenty persons who were given a course on professional management. This turned out to be the best and most comprehensive management course that I have ever been exposed to. The course made a lot of sense. Lights were turning on for me much of the time.

One concept, however, did not go down well with me or the other engineers in the group. This concept was that a manager could be moved into any position regardless of his technical expertise. "Do you mean to tell me that an accountant in charge of the financial function could take over as the manager of engineering?" was my challenging question, for which I got the reply: "Yes. If a manager is a professional and knows how to manage people, it doesn't matter what they're doing. A manager can be a general purpose manager and move to any position in the company." This raised the hackles on my back. It remained a challenge for many years. It just did not sit well.

Some theories and concepts do not stand the test of time and the "general purpose manager" concept was one that failed in this large organization. They tried interchanging managers from finance, engineering, personnel and marketing for about five years, then licked their wounds when it did not work. Now, many years later, as a management consultant well-versed in the engineering design process, I know why there is a difference for the management of engineering design.

Nothing I say in this chapter should deter you from learning all there is to learn about general purpose management. You must know how to plan, organize, delegate, follow up and integrate, just as I was taught in my professional management course. Any manager worth his salt learns to do all of these things and to do them well. In addition, the engineering manager must learn a few extra things. That's what this chapter is about.

WHAT THIS CHAPTER WILL DO FOR YOU

You will learn that there are five reasons for the engineering manager's job being different from that of other functional managers, then you will find out what to do about it. You will learn about "granted authority," about separating the "what" and the "how," and about how to nurture creative talent. Then, the part on "a day in the life of the engineering manager" pulls it all together. Finally, there are specific Do's and Don'ts for engineering managers.

WHY IS ENGINEERING MANAGEMENT DIFFERENT?

1. Loss of Expertise

At the time of his/her promotion to the first ranks of management, an engineer is a technical expert on something. As a new engineering manager, he/she can call the shots by virtue of the superior technical expertise which was developed by years of experience and specialization. However, in the course of time, the engineering manager will get to know more and more about the business aspects of his/her job, and less and less about the technical aspects. He/she may cart home a lot of reading to do at night and pore through journal after journal, reading the state of the art technology. He/she may enroll in refresher courses in his/her technology. It is right that an engineering manager should do these things, but there is *no way that he/she can become an expert in all of the technology involved,* particularly if it is a fast-moving technology. The engineering manager is a loser when it comes to keeping abreast of technology. His/her subordinates work all day on a specialized part of that technology. They get their knowledge from experience, from talking to other experts, and by reading in the journals. The fine points of a technology are learned by the people who practice it. There is no way the engineering manager is going to keep up with the efforts of five or ten other people, each specializing in one part of the common technology. Whatever *earned respect* that the engineering manager had on promotion will soon be lost if based only on technical expertise. It can and must be replaced by earned respect for his effectiveness in management.

Contrast this job with that of the sales manager. If this person once had a track record as a salesman, he hasn't lost the touch. He continues to practice his techniques of selling and continues to maintain the respect of his subordinate salesmen because of his knowledge about selling. Whatever changes come about in selling techniques, they are easily picked up by the sales manager. He does not face rapid technological obsolescence as does the engineering manager. See Figure 12.1.

2. Source of Major Decisions

In classical management theory, information flows upwards through channels in the organization, and decisions flow downwards. This is logical when the resources must be controlled by those at the top and they have the expertise to do it. This is not true in engineering design, especially where a high technology is involved. The decisions that really count are made by those with the technical expertise— by the engineers near the bottom of the hierarchy. These performance decisions have a far-reaching effect on the time and cost of

ENGINEERING MANAGER	SALES MANAGER
Comes up through the ranks because of his superior technical expertise.	*Comes from the ranks of salesmen because of his selling expertise, which includes the ability to get along with people.*
Inevitably loses the superiority of his technical expertise, which was the basis for the engineers granting him authority to coordinate. He becomes technologically obsolescent.	*Continues to have the know-how about selling and getting along with people. Usually participates in the selling.*
Needs to get a track record for his skills in coordinating and communicating in order to get granted authority from the engineers.	*Manages mainly people like himself, who have a similar know-how about selling.*
Management style must change according to the design phase being at the beginning or end of the design process.	*Management style strongly persuasive for most of the time.*

Figure 12.1. Differences for Engineering Managers.

a project. Top management is only in the position of controlling the effects of the decisions rather than initiating the decisions which really count.

3. The World View of the Engineer is Different

A "we" and "they" attitude exists between engineers in design, and corporate executives. The difference in their views of the world about them is no accident. Engineers are a product of an education which was very much influenced by the philosopher Leibniz. They believe that the world is orderly and predictable and can be modelled. For example, Newton's laws of motion are dependable and can be used for making predictions about the results of an engineered design. The world can be manipulated, it does not control us.

The world view of the corporate heads is more closely associated with that of economists who are associated with the philosophical school thought arising from Locke. They take a pragmatic and compromising view of the world. Survival of the corporate entity is far more important than the sponsorship of new ideas. The world rules us, we should be practical and conform.

The creative bent of engineers is continually tempered by the financial decisions of those in top management, who, not surprisingly, think that engineers have their heads in the clouds. It is the job of the engineering manager to bridge this gap by his communication skills. This means that he must be able to share and reconcile widely different world views. This makes his job different from that of, say, the financial manager of an organization. (See note 1.)

4. Coordination with the Nontechnical Units

No engineering design can become a reality without the work of many nontechnical persons. Sales, personnel, finance, and production must all do their part for the design to be really successful. The engineering manager must coordinate the efforts of his engineers and scientists with these nontechnical units within the organization. This is a horizontal type coordination and not the same as the up and down communication of different world views mentioned above. The point is that nonengineering functions tend to become separate entities which can operate more or less independently for most of their work, whereas no engineering design can become a successful reality without the coordination of many persons who are not engineers. Thus, the engineering manager is more dependent on other functions than they are on him.

5. Flexibility in Management Style

The engineering manager's relationship with his subordinates will not only be determined by all those things which normally make good manager-to-subordinate relationships, it will also depend upon where they are in the design process. For this reason, the engineering manager's style may range from laissez faire to participative to authoritative, depending on where they are in the design process. Indeed, if there are many designs under way, the style may change many times during the day. This is a very important difference for the engineering manager to comprehend. It is dealt with separately below.

THE ENGINEERING MANAGER MUST HAVE FLEXIBILITY IN STYLE

As an engineering manager, you should know how to manage *with* the engineering design process and not *against it*. When you understand what design is all about, you will vary your management style to suit the situation.

Consider, for example, the beginning of the design process. In the early *concept phase,* your management style should be laissez faire because your engineers need the freedom to create. In the development phase, there is a fair amount of time and still not much money invested when compared to the final completion time and cost. You can work together in a problem solving style with plenty of participation. Most of the decisions at the early stage are technical and can be made jointly with your subordinate engineers.

However, consider a design which had progressed to the *detailed design phase* and is being prepared for assembly or production. The design is almost frozen or, as some say, "set in concrete." It is costly to make changes. The pressures on time and cost are strong, particularly those pressures on you, the engineering manager. The test is on. Is the engineering design going to be successful? Success means being on time, being producible, and at a reasonable cost. Now you are called upon to make more of the decisions. *You must "call the shots," so to speak, even when these decisions are unpopular.* It is always unpopular to end the development on a design. Your design engineers want to continue to make it better. Literally speaking, they never know when it's finished. You, as a pragmatic engineering manager, know that 99 percent is good enough. Your design team must get something finished or you and they will never get another job to do. During this period, human relations are likely to be strained. You are likely to be authoritative in many of the decisions. Unless your engineers have granted you the authority to do this—you could be in trouble. See Figure 12.2.

GRANTED

When you were first promoted into engineering management, you may have been a technical expert with the earned respect of your peers. They may have "granted authority" to you to make the key technical decisions. However, as I pointed out, sooner or later they will make the major decisions because of their technical nature.

Under what conditions will your subordinates allow you to make major decisions about time, cost, and performance when the locus of expertise is with them? They will do this *when you have a good track record and you have their earned respect through your expertise as a manager.* Without doubt, they need you to communicate with upper management and to coordinate their efforts with the nontechnical parts of the organization. They need you in order to have final success with an engineered design. If you must be authoritative at the end of the design process, and they believe you will lead them to success, they will grant you the authority (See Figure 12.3).

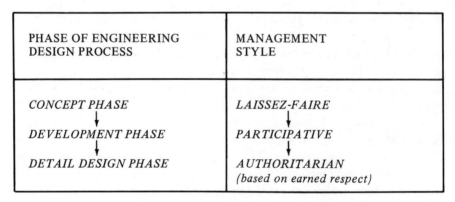

Figure 12.2. Management Style Changes.

BASIS OF GRANTED AUTHORITY TO COORDINATE	GRANTED BY ENGINEERS
EXPERTISE (KNOW HOW)	*When the subordinate engineers recognize that the engineering manager has the technical know-how to make the decisions. He inevitably loses this basis for authority.*
EARNED RESPECT	*When the engineering manager has demonstrated that he can coordinate and communicate with the skill that leads to repeated successes, the subordinate engineers follow him in difficult situations. At other times, he gives them professional freedom to exercise their technical know-how.*

Figure 12.3. Granted Authority.

Notice that I have used the words "granted authority," as if your authority did not come from above. *"Granted authority" is that willingness to cooperate with you when you make unpopular decisions.* Your subordinates are mostly professionals and to a large degree they themselves decide on the details of what should be done. When *you* decide on the details of what is to be done, you are encroaching on their prerogative of determining the *how.* You do this with caution. It helps if you have developed a *charisma* by chalking up good track records with other design projects. Then they will really believe that you will lead them to success once again. You must, therefore, be a top-notch manager in every respect. You replace your technical expertise with management expertise. You show by its exercise that you know what you are doing (See Figure 12.3).

An engineering manager can easily learn those advanced techniques of management arising out of management science. You can use quantitative methods in group decision making which allow for inputs from all levels. You can understand and use modelling techniques which will be useful in other branches of the organization. In other words, you are able to show managerial leadership in those areas which dovetail with your engineering knowledge. All of this is in addition to your necessary skills in the ordinary aspects of management.

THE SEPARATION OF THE WHAT AND THE HOW

If the engineering manager and his subordinate design engineers are to truly share the responsibility for the success of the design, a separation of the *what* and the *how* of the design will help keep their responsibilities clear and separate. The professional has a technical expertise and really knows best "how" to make something work. On the other hand, the engineering manager has a connection with the sponsor of the design and surely must have some say on the *what,* particularly on the timing and cost of the design project. Yet, because performance, cost, and time are so interrelated, the engineering manager definitely gets involved in the performance. It is important to separate "what is to be achieved" by the performance, from "how it is to be done."

Fortunately, the performance variables can be separated into the *what* variables and the *how* variables, as was shown in chapter three on Objectives and also in chapter ten on Optimization. The engineered design creates a new state in which certain variables are changed to new values as outputs. These are the *what* of the design. They correspond to the outputs of the "black box." It is clearly these that the sponsor, customer, or client wants from the engineered design. On the other hand, the professional design engineer is very much concerned with what goes on within that black box. These we have called the "new design variables." They are engineering details which are best left to the design engineers themselves. Thus, the smart engineering manager will keep his fingers out of the design details and put his emphasis on the *what* variables which are mainly the inputs, outputs, and constraints.

There are two situations that the engineering manager should be aware of in separating the *what* and the *how:*

1. When dealing with a subordinate, it is convenient to separate the *what* and the *how* in the discussions about design parameters. This minimizes any conflict which may arise. As engineering manager, top management holds you responsible for the *how* as well as the *what.* In other words, you are responsible for the quality of the designing generated in your group. For this, you review, promote and otherwise reward your design engineers.

When you are reviewing and evaluating the performance of subordinates, the *what* and the *how* come together. If you are lacking in technical expertise, you may need to get some advice about the quality of the *how* from the technical peers of your subordinates.

Do not assume that you will always have enough technical expertise to judge the quality of the *how* by yourself.

2. If you are a project manager as well, then you are responsible to upper management for the *what* of the engineered design. If the person doing the engineering does not report directly to you as a subordinate (as in matrix reporting), then he/she and his/her functional manager are responsible for the *how.* Keeping these two things separate will help you avoid having unproductive conflicts with his/her functional manager, who, by the way, will often be another engineering manager. When you happen to be both an engineering manager and a project manager, you should be capable of operating in the above two modes. For some tasked persons, you are also boss, for nonsubordinates, you are a project manager with authority only for the *what.*

EXAMPLES OF SEPARATING WHAT AND HOW IN MATRIX REPORTING

What and How for a Gear Box

Project manager to nonsubordinate tasked engineer: We need a gear box with a 3 to 1 ratio and a torque capability on the input of 100 ft. lb. (*what*). I think you can use low-cost spur gears from castings (*how*). (Other *what's* are r.p.m., interface dimensions, efficiency, performance tests, life tests, means of lubrication.)

Tasked mechanical engineer to project manager: What are the shaft sizes to be (*what*)? Thanks for your suggestion on the gears. I'll take it up with my boss (*how*). (Other *how's* are bearing choice, materials, lubricant, fastenings.)

Boss to tasked engineer: The shock conditions are too severe for a cast spur gear. You must use machined bevelled gears. You can off-set that cost by using a sleeve bearing. (*How*).

Project manager to subordinate engineer: This is what we are using for the drive motor. Design a belt drive to connect it to the gear box (*what*). Use a double vee-belt (*how*). Let me check the drawing details before it is finalized (*how*). (Other *how's* are spacing of belts, pulley size, clearances.)

What and How for a Maintenance Manual

Project manager to nonsubordinate tasked person: Your boss has assigned you to this task (*how*). What we need is a manual that a maintenance man with 6th grade reading skills can understand (*what*). They can't be expected to understand our blueprints (*what restraints*). Do you have a way to make this gear box easy to understand (*how*)? In addition, it needs to have a waterproof cover, be easily carried, and cost us less than $5 each in quantities of 500. I think you should use bold gothic type (*how*). (Other *what's* are maintenance procedure and frequency, replacement parts to be listed, limits on substitutions, use of company logo.)

Tasked manual specialist to project manager: We can have exploded drawing prepared, but not within your cost target. Do you want understandability or low cost? I don't know how I can do it. Anyway, don't concern yourself with the typestyle. That's a detail that I can look after (*how*). (Other *how's* are the graphics design, layout, order of sections, paper grade, binding.)

Boss to tasked person: Exploded drawings would be too expensive. There is a new technique of touching up photographs and that should do it. Look into it (*how*).

Engineering managers who are also project managers will do well to study Figure 12.4 and see where they fit. The separation of the

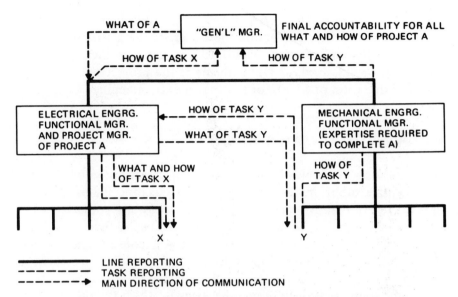

LINE REPORTING
TASK REPORTING
MAIN DIRECTION OF COMMUNICATION

WHAT = WHAT IS TO BE ACHIEVED BY THE TASK (INCL. RESTRAINTS IMPOSED BY THE INTERFACE WITH OTHER TASKS – WHICH APPEAR TO BE "HOWS")

HOW = HOW THE TASK IS TO BE DONE (DEFINITELY THE DETAILS OF VARIABLES THAT DO NOT DIRECTLY DEFINE INPUTS, OUTPUTS, OR INTERFACES)

Figure 12.4. Matrix Reporting When Project Manager is Also a Functional (Engineering) Manager.

what and *how* in practice is the way they have teamwork. People are treated as professionals. Unproductive conflict is minimized.

What and How for a Construction Project

Project manager to a nonsubordinate construction foreman: According to the specs the excavation for the footings is to bed rock, supposed to be 16 feet down (*what*). I think you'll need a large pump to keep it dry (*how*). Can you have it finished before Friday? (*what time*).

Tasked construction foreman to project manager: We'll be ok because I did that building over there and there's no ground water here to worry about. I'll get a small pump if it rains (*how*). Friday? The schedule allows us to next Tuesday and I've arranged for machines accordingly. I'll see my boss about getting it done sooner.

Boss construction superintendent to tasked foreman: The project manager could be right about the water. Leave a partition in the trench at the end of every wall until you are sure there will be no flooding (*how*). Friday? OK with me if the project manager signs an authorization of $500. for renting an extra machine (*what cost*).

NURTURING YOUR CREATIVE TALENT

The separation of the *what* and the *how* is necessary in your relations with creative subordinates, as well as with nonsubordinates. It stimulates the use of creative talent in all people, including yourself.

Let us take a look at what truly creative people are like:

First, they (or you) are open to experience. Inconsistencies in technical relationships do not upset them. Instead, they see them as challenges which can be mastered and explained.

Second, they (or you) are creative because they are able to work with the preconscious mind which is teeming with ideas and fantasies. Recent analysis of brain surgery results indicates that in right-handed people, the right side of the brain is the creative part and the left side is the judgmental part. For creative ideas to come into the conscious mind they must pass the screening of the judgmental part of the brain. In some people, this comes naturally. In others, it can be stimulated by first generating ideas in a freewheeling fashion, writing them all down, and then judging them (See chapter six). Some people have "hardening of the arteries" of communication between the left and right side of the brain. They are not very imaginative, but may make up for it in judgmental powers.

Third, the locus of satisfaction is often within the mind of the highly creative person. They don't care what you think of their ideas. This makes it tough to be the boss of a creative person, but you can handle it when you know the reason behind it. Imagine yourself evaluating paintings by avant garde artists. Some of them don't care what you think of their creations. In their view, you are artistically dumb. So, when you turn down a "wild" idea from a highly creative person, you have not knocked their idea down, not really, you just don't know any better.

Personally, I have found it stimulating to discuss creative ideas with a highly creative person. If I can't adopt each and every great idea, they understand my position. They know that I appreciate the "idea," even when I can't use it.

Fourth, and not unrelated to the above, they often will not work on problems with routine solutions, or, not for long, and they are not happy doing work which can be readily evaluated by others. When a job loses the luster of innovation, they are likely to seek another job.

Fifth, sometimes fear motivates them, but not always. Is the creative genius afraid of losing his job? Sometimes, but not always. They are seldom motivated by your praise, or a raise, or the threat of losing their jobs. Real oddballs. What, then, is left as a motivator? Not much, because they are already internally motivated to solve tough problems—so just keep flinging challenges their way.

Sixth, creative people have an ability to conceptualize. They can take two ordinary things and mentally or visually rearrange them in a new order which was never shown to them. This is probably their innate talent. You can only stimulate it or repress it. Be glad that some people can conceptualize better than others.

Seventh, creative people are not always "nonconformists" in their manners and dress. True, they do not conform in some part of their thinking, which if you are lucky can be in the technical area. On the other hand, if they don't care about manners and dress, this is not teen-age insubordination. It is a characteristic you might get with a near genius. Engineering managers who must have innovative subordinates, usually say that they can tolerate the "problem areas" if the ideas are really good.

Now, if you are the creative person, cool it on the nonconformity that disturbs other people. Most engineered designs require teamwork, and people who appear to fit the mold. If you are in command of your creative talent, you can turn your nonconformist thinking to the technical problems when you want to and conform a bit in other matters when it helps teamwork. Find a *what* that challenges you and create the best *how* that you can.

In summary, creative talent is better nurtured than repressed, whether it be yours or someone else's.

In nearly all situations, you can come out ahead by separating "what is to be achieved" from "how it is to be achieved."

A DAY IN THE LIFE OF THE
ENGINEERING MANAGER

Below is a description of one working day of an engineering manager. This is neither typical nor atypical. It is, however, based on the real experiences of myself and others.

John is an engineering manager for Apex products. He is responsible for the design of consumer products, but occasionally he takes on special projects such as the design of a new factory. It is Friday morning. John enters his office at 8:30 a.m. and finds a big note sitting on his desk, "Call the production manager at once."

While John was on his way to work, the factory had already been working an hour and had uncovered a problem which they think is the fault of engineering. The production line is about to be shut down and it will cost $500 an hour if John doesn't do something about it. He is reluctant to break into the important work of one of his talented development engineers who is working on a brand new idea for a product. However, he has no choice. This development engineer, Joe, is the designer of the product now in production—although he did it all of two years ago. He is the only person with an intimate knowledge of the product. He can jump in there and find out how to fix it. John approaches Joe with caution and apologizes for interrupting his work. He describes the *what* of the problem and directs him to go and see the production manager immediately. For John there is no choice. For Joe there is no choice either. He hopes there will be a challenging problem but he thinks this is just going to be another waste of time.

While Joe is away solving the production problem, John gets ready for an important interview scheduled for 9 o'clock. This is a vital one. He is trying to lure away a highly creative engineer from one of his competitors.

The interviewee arrives thirty minutes late and doesn't apologize for it. He is casually dressed in patched jeans and open shirt. While talking to John, he pulls out a cigarette and lights up without an invitation to smoke. John knows that if the personnel department had interviewed this person, the interviewee would never have gotten this far. They like conformists. But John knows that this man can really produce the goods and he is not going to judge him on his manners and dress. John finds out that it is not money that this person wants. Instead, he would like to have the freedom to come and go as he pleases—to work long hours if he wishes, or to take leave if he wishes to. He doesn't want or care about a steady pay check.

So much for the interview. It's over in an hour. John will have to deal with his personnel and financial management before he can talk further to this prospect. As a matter of fact, he hasn't much hope for changing the way the engineers are paid. The executives like to run a tight ship with everything going smoothly. It's not practical to have exceptions. On the other hand, they also like to have profitable new ideas which keep them a step ahead of their competition. John sits in the middle and must somehow bridge the gap between the world view of the creative engineer, and the world view of the personnel and financial managers who like to maintain the status quo. As John sees it, his only hope is to bring the creative engineer in as a contractor and not as an employee. Maybe it will do the trick.

Eleven o'clock rolls around and John takes a few minutes to check on the production problem. It turned out to be a vendor's component and not the fault of engineering. He listens to his engineer's gripe—and feels it as his own.

Next, John prepares for the weekly meeting with his boss and fellow managers. He puts on his suit coat, being consciously aware of the different personal values about dress which are informally known to the executives and to the engineers. John doesn't find this strange because he has been studying sociology in the evenings. He is prepared to conform in manners and dress because it's no big deal for him. He has learned something about financial accounting so that he can communicate better with the comptroller. He is thinking of a course on personnel administration but is not yet prepared to admit to himself that he needs this to bridge the communications gap between himself and the personnel manager.

When John first joined the executive team he had some friction with the managers of the nontechnical functions, but gradually he was able to see their point of view. He hopes he can educate them to see his point of view too—such as being called into a production problem that was a waste of his engineer's time, and having to deal with special types of employment arrangements.

It is 2 p.m. The executive meeting lasted longer than usual and lunch was brought in. Now John is conducting a project review meeting with his engineers. Representatives from the marketing, the production and the service departments are present to review the results of testing the first prototype of a new product. The service department thinks that access to the product should be simplified but the production department doesn't want any complications in their assembly. Eventually, things work out quite well. They are able to put their heads together in a participative effort at jointly solving the problem. They start out by defining what is to be achieved. Suggestions on the *how* is made by all parties, but the final decision on the *how* is up to his project manager. John enjoys this kind of meeting because he can exercise his skill as a problem solver and also gets the chance to promote teamwork between himself, his engineers and other parts of the organization. At this stage in the design process of the product under review, the problems are not very technical. He can participate without trying to play the role of an expert. He gave that up a long time ago. Instead, he is aiming at becoming a top-notch manager by directing his continuing education to learning more about those subjects which the art students take at the university—the neglected part of his education. They will help him to be a better manager.

Author's note: What I have shown you here is that the day of an engineering manager can vary from a downstream design activity where an authoritative style is necessary, to upstream activities where his style is either participative or laissez faire.

Moreover, the engineering manager is bridging the social, cultural, and technical gaps between his engineering team and other parts of the organization. He does this with the tact and skill which comes from knowing what management is all about. He does it even better when he *understand the difference* between his job and the job of the other managers.

SUCCESSFUL ENGINEERING MANAGEMENT

In this chapter, you discovered that the engineering manager's job is a bit different from that of other functional managers. Because of the loss of technical expertise, technical decisions flowing upwards, the view of the world as predictable, the need to work with non-technical units, and the need to be flexible in management style, the engineering manager must obtain "granted authority." This is done by demonstrating skill in communicating to the nontechnical management, nurturing creative talent, separating the "what" and the "how," and knowing when to be laissez faire or directive.

If you wish to become a cut above average as an engineering manager, practice the following Do's and Don't's and follow through with the exercises at the end of this chapter.

GUIDELINES FOR ENGINEERING MANAGERS.	
Do	**Don't**
Do realize that many non-technical persons do not share your view of the world.	*Don't get tagged with an "engineer's mind." Not everything is orderly and predictable. Other people have other ways to look at a problem.*
Do simplify your technical jargon when communicating upwards.	*Don't be an egg-head. You may get cracked open.*
Do try to keep up to date technically in breadth of knowledge, but not in detail.	*Don't try to be a know-it-all boss. Eventually, you will slip behind in your technical specialty. When you realize this, your subordinates will cheer.*
Do learn about all aspects of management. Learn all you can about management science too.	*Don't say you haven't time for management training. You must, or else join the large club of engineering managers who don't know beans about management.*
Do practice flexibility in management style, depending on where you are in the design process. Allow for people variations too.	*Don't adopt a single management style because it is popular. Learn to be easy-going while upstream in the design process. Change to tough and directive when you need to finish up a project.*

Do nurture the creative talent of yourself and others by encouraging ideas. Give them some freedom in lifestyle too.	*Don't waste time laughing at an idea from a genius. He or she doesn't care about your opinion.*
Do separate the what *and the* how. *Set clear objectives with measurable criteria based on your knowledge of the requirements.*	*Don't get your fingers in the details of the* how. *You might get burned. Let your subordinates determine the* how—*less burning all around.*

Questions to Consolidate Your Understanding of This Chapter

1. What are four reasons why the design engineering manager's job is different from that of functional managers of mainly routine operations such as finance, production, and sales?
2. Why must "granted authority" come from persons below instead of from above?
3. How does the engineering manager obtain "granted authority" from professional subordinates?
4. If at the end of a design project, the engineering manager finds it necessary to become authoritarian, under what conditions will the professional subordinates accept his directives?
5. In what ways does the world view of an engineer tend to be different from those persons educated in arts or social sciences?

Questions on "A Day in the Life of the Engineering Manager"

1. John is reluctant to break into the important development work of Joe. What are some alternative actions that John might have taken? Make a short list, and for each tabulate the advantages and disadvantages. Make a recommendation and state the conditions under which it would be better than what John actually did.
2. John wants to hire a highly creative engineer but the conditions of employment will be difficult for the employer. Assuming

John needs some creative talent that he doesn't already have on staff, what are some alternative strategies he could try?

3. Consider the product review meeting that John conducts. If, instead of reviewing the first prototype, they were reviewing pilot run results and full production was to start in a week, what would be likely to be different in John's management style during the conduct of the meeting?

Assignments to Help You Become
A Top-notch Engineering Manager

1. What percentage of engineering managers that you are acquainted with have actually taken training in management of at least three full weeks or its equivalent part time? Try to get similar data on engineering managers who were not successful.

2. Make a list of educational opportunities open to you for learning about planning, organization, delegation, interviewing, appraisal, interpersonal communication, group dynamics, and other basics of management know-how. If you haven't covered them, make a plan to do so.

3. Make a list of the engineering projects that you are working on. Classify them as currently upstream, midstream, or downstream in the design process. For each, indicate whether your management style should be generally laissez faire, participative, or authoritarian. Now go into action and practice what you have learned.

Note 1: Dr. Myron Tribus of the M.I.T. Center for Advanced Engineering Study talks about the engineer's work view in an article in the IEEE *Spectrum,* April 1978. The contrast with lawyers is enlightening. It appears that we engineers are the servant problem solvers instead of being masters who decide what problems should be solved.

Dr. Ian I. Mitroff and Dr. Murray Turoff reveal how much our basic assumptions about the nature of the world affect our approach to solving problems in the IEEE *Spectrum,* March 1973.

13 Computer Aided Design: CAD Enhances Your Capacity

KEY IDEA: COMPUTER AIDED DESIGN(CAD) AND COMPUTER AIDED MANUFACTURING (CAD/CAM) TOOLS WORK. DESIGNERS AND ENGINEERS WHO CAN APPLY THEM ARE WORTH MORE.

This chapter contributed by Dave Hogg of the Ontario CAD/CAM center

OBJECTIVE
You should gain an understanding of the nature of CAD–CAD/CAM technology, its impact, and some of the basic rules to apply when investigating its application to your need.

BENEFIT
You can lever your present capabilities by an order of magnitude or more through the proper application of computer aided design and manufacturing tools.

This chapter will put CAD into an easy-to-understand perspective. It provides a framework which will enable you to keep your thinking straight as you begin to develop an understanding of how this technology may be applied to your

238

particular application.

There is no longer any doubt. Computer aided design tools work, and work exceedingly well. In fact, they are powerful productivity weapons in the hands of designers and engineers who know how to apply them. A simple survey of the North American balance of trade deficit in many manufactured goods drives the point home.

Although the new tools are "different" in many respects, they will help you do what you do now - but do more of it better, cheaper, and more easily. However, it's these differences that account for the fact that the vast majority of potential users have yet to adopt them.

Why the hesitancy? Quite frankly, these tools are new to North American managers who have yet to acquire the confidence needed to be able to apply these tools effectively. Technically, the tools are proven and perform well. But as we wait, it is becoming increasingly clear that future markets will go to the countries, and the companies, who will make the most intelligent use of these new productivity tools which are the first substantive products of the information age after the microchip.

The very way in which we make things is changing literally before our eyes, and changing forever. Unfortunately too many of us are spectators watching the change - and not participants.

Consider Bill W.'s small electronics firm which reduced its printed circuit design time from 4 weeks to 4 days for its bread and butter line. This single CAD benefit in only one process provided Bill's firm with the time needed to hold off the competition while even more advanced tools could be investigated, justified and purchased to reduce the total product design and delivery cycle substantially. They are surviving today and will so tomorrow.

The examples are everywhere, but don't be mislead - not all experiences are as positive. To put things in perspective, a report by A. D. Little, a prestigious consulting firm, indicated some 20-50% of the installed CAD/CAM systems failed to meet objectives. Like any new tool which is based on a technology that is new to its user, a period of adjustment is needed in order to build the required confidence which precedes any significant investment decision. A considerable number of impulse acquisitions have occurred and failed because of a lack of understanding of the technology and the need for training.

SO WHAT IS DIFFERENT ABOUT CAD/CAM TECHNOLOGY?

Most differences are imagined. Those who do not understand computers (and that is many of us) accord them a mystique that's not justified. In reality, approaches to justifying, selecting and applying the technology still follow pure common sense – once a clear understanding how these new tools work is acquired.

CAD/CAM systems are based on computers – tools which only in the past decade have earned respectability in manufacturing. In fact they have yet to earn credibility with the majority of CEO's and senior engineers who make the key adopt-or-not-to-adopt decisions. Unlike the simple chisel or even the standard milling machine whose outputs are easily calculated and understood, tools such as CAD/CAM, require a solid knowledge base in order to "grab their handles". No matter what the salesperson says.

Computers require accurate and timely information that is properly formatted. **And since many companies do not have precisely accurate information in the needed form, preparation time must be spent to adapt to these new tools.** It is in this regard that many implementations stumble. The basic concepts of data backup, software and hardware maintenance, and system security are initially not well understood. Many have difficulty understanding why they should pay a software maintenance fee to correct bugs present in the software they purchased. Such are the uncertainties with this new technology. These will pass.

SO WHY CONSIDER COMPUTER BASED TOOLS?

As a designer, engineer, or project manager - your future will be largely dependent upon knowing what these tools can, and cannot, do and how they can be used. Quite frankly, the leverage these tools provide will force firms out of business who can't see their value. It's happening already. Don't underestimate the impact of the policies being adopted by the major automakers. They are saying quite bluntly to many of their suppliers that those who cannot handle digital data will not be allowed to quote on new jobs. In this case technology is being driven from the top down and a supplier has little choice it he wishes to stay in business.

The impact is being felt in all sectors in countless ways. To get the point across, one consultant refers to CAD systems as the never-to-have-to-draw-anything-twice systems. An overly simple concept but a powerful one. Imagine how one could lever this one aspect alone in doing what they do. Think of the time that can be saved by calling up a part(drawn only once) from a memory bank. A part that is reproduced in a tiny fraction of the time with 100% accuracy - every time! Think of the impact it could have on establishing standards across a company and the savings that it would yield. But these are only a few of the benefits the technology offers. These new tools of the information age are **amplifiers of cybernetic feedback in the design process** as described in Chapter 1.

> A large western consulting firm obtained a job to size and cost the pilings needed to support some 23 acres of decking for a theme park.

The firm's civil group were excited because it meant approximately 2-3 weeks of work for two skilled engineers. However the firm's interactive graphics group knew a better way. Having much of the data already in the data base from a similar job, the manager had his CAD operator sit at the computer for one day. She fully sized and costed all 11,518 pilings. Needless to say the civil group was not pleased but from a corporate perspective the firm made a good profit on the job. Such are the impacts of this technology on certain tasks.

Progressive small and medium size firms are beginning to use these new tools to gain an advantage over their competition, and especially to enable them to compete with larger firms who have difficulty in responding quickly to change.

As the recession was drawing to a close in the early '80's, the president of a small job shop in northern Michigan supplying automotive parts became excited with CAD/CAM's potential. Based only on his personal belief, the firm spent over half a million dollars to ride the 'new wave' as he called it. Like most pioneers they initially lost money and experienced many startup problems. However, through his personal commitment and an excellent sense of timing, the firm more than doubled its staff while quadrupling their gross income. His investment paid off — for while others were shaking off the effects of recession, he was expanding his market share with a technology he understood an could apply.

Because of the speed of technological change today, there is no time to recover for the company who belatedly recognizes the leverage and benefits available technology offers. In a seminar to manufacturing executives, Edwards W. Demming, the man many say is most responsible for Japan's current manufacturing position, expressed the view to the effect that "one can not expect to ever catch a competitor by copying him - your only hope is to do an end run and find a better way of applying the new technology to your advantage". And that will require a level of commitment many will find exceedingly demanding.

Competitive position and advantage are everything - and will be so to the end of the decade. John Naisbitt in his book "Re-Inventing The Corporation" states that over the past 5 years 40% of the new net jobs came from companies that did not exist 5 years ago.

Unsettling statements are being made by such prestigious groups as the National Science Foundation's Centre For Productivity who have stated that "CAD/CAM has more potential to radically increase productivity than any development since electricity". Despite how extravagant that statement seems, it appears more and more plausible as one sees what the new technology is capable of... when properly planned for and implemented.

In a nutshell, it appears that while implementation and application of advanced manufacturing is sputtering in North America, solid expansion continues unabated in the countries of the Pacific Rim. This development will no doubt force North American manufacturers to pick up the pace in order to successfully ride the new wave into the future.

WHAT IS CAD/CAM?

Consider for a moment, a "knowledge tool" that never forgets, that eliminates all that routine number crunching, that enables you to produce almost error free drawings and designs, that keeps your best designs at your finger tips, that maintains scrupulous standards, and on top of all this - gets more accurate as time goes on. Spark any interest?

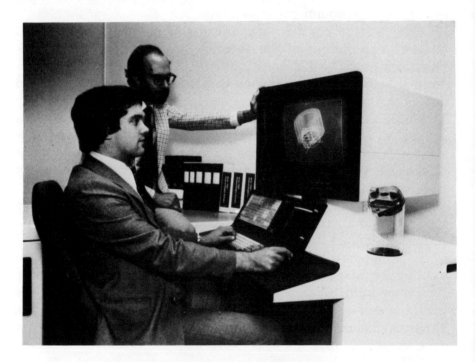

Figure 1:   A typical CAD workstation showing the operator entering data while observing the developing model on the workstation's monitor. Instead of the physical act of drawing arcs, lines, etc. the operator issues commands which the computer executes and stores. Repetition is removed as parts seldom have to be drawn twice.

Very simply CAD stands for Computer Aided Design.. the use of the computer to automate the functions needed to design a component, system, or product. The first CAD systems began to appear in the early 1960's but were really only computerized drafting systems. Computer Aided Manufacturing(CAM) only developed significantly after the second world war - and with the advent of 'chips' from the space program it literally took off. The heart of CAM is basic Numerical Control(NC) technology, which utilizes microcomputer-chip logic to drive machine tools of all kinds.... today from a paper or magnetic tape, but tomorrow - directly from the CAD system that developed the digital design data in the first place.

As with any new technology, different terminologies abound which generate initial confusion. To keep the terms in perspective, visualize the factory of the future as an entity wherein all the processes and functions are computerized and in turn "orchestrated", by an executive computer. This is an example of Computer Integrated Manufacturing or "CIM" for short - a term we will all be seeing more of to describe the "factory of the future". Into such a factory would flow market information, ideas, and raw materials with finished products flowing out.

CAD/CAM is simply a component of CIM. Robotics, factory automation, manufacturing information systems, automated warehousing, and inventory control are also components of CIM.

CAD receives most attention simply because it is easy to understand and more companies can use CAD more quickly - usually beginning with drafting applications. It's not new. CAD/CAM has been with us since the mid 1970's. What is new is the and sheer computing power of today's machines, and the ability to now transfer data directly from the CAD system to the CAM machines that make or test the part, system or product. The 'slash' between CAD and CAM will disappear as this transition increases.

Those in the Architectural, Engineering, and Construction fields tend to use the term CADD which better suits their application.   It stands for Computer Aided Design and Drafting(CADD).   In process industries such as pulp and paper, you will not hear the word CAD/CAM - rather 'mill wide control' even though many of the concepts and even much of the hardware used is the same.

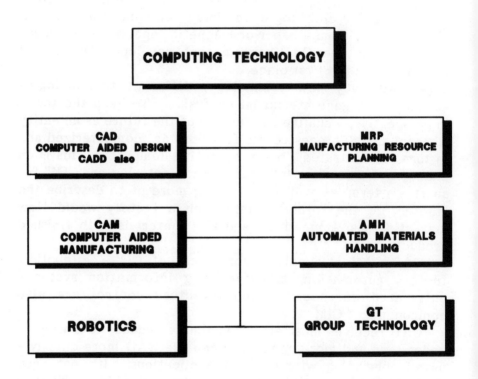

Figure 2:   The key components within Computer Integrated
Manufacturing

SHARPENING THE PERSPECTIVE

As has been pointed out, the vast majority of North American firms do not have CIM, but the concept is becoming increasingly applied as a planning perspective, a corporate strategy, and a state-of-mind for planners and managers as they begin to purchase CAD and CAM systems. Already the pursuit of methods for linking these individual components together is underway. CAD/CAM is really the wheels that will propel many firms into the world of CIM.

CIM is the framework around which the technology will evolve - logically and productively.

Typically, one's experience might go something like this.

* A corporate strategy is ratified with a CIM or Technology Action Plan developed - and communicated to all as part of the business plan.
* Corporate champion appointed.
* CAD, CADD, or CAD/CAM investigated to reduce design cycle time, reduce shop errors, improve product quality, cut design and manufacturing costs and so on.
* Pilot project to get 'biggest bang for the buck' is defined, funded and scheduled.
* Capacity increases, marketing/sales efforts expanded.
* New ways of exploiting the systems are discovered.
* Pressure for unplanned system expansion grows.
* Means of linking the 'islands of automation' investigated.
* CIM plan constantly reviewed.

IMPLEMENTATION DECISION FLOWCHART

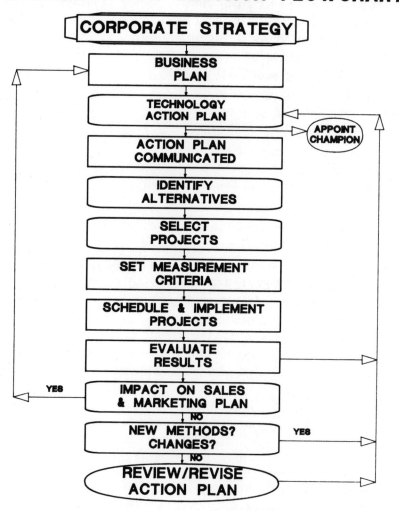

Figure 3:   Flow chart showing typical implementation decision
sequences for new implementers of advanced
manufacturing technology. Note the similarity to the design
steps shown in Figure 1.2 of chapter 1.

One of the more hidden benefits is the inherent development of a common data base for all to share. No more worrying if the drawing you have is the latest - if it's from the system it is the latest. As this data base develops so does its intrinsic value to the company. Opportunities will arise to automate the inventory control system, to provide for the automatic order of stock and raw materials - and so it goes. Factory scheduling, process planning and all that it entails becomes another 'add-on' to a system that begins to truly integrate all the functions and processes that go on in a typical plant. But it will take much time and expense.

Today, islands of technology are springing up in firms of every size. CAD is purchased to automate the drafting function, to design printed circuit boards, to analyze the thermal and mechanical stresses on a part before it is made. CAM for many small firms began with the purchase of a Numerically Controlled (NC) machine which invariably leads to interactive programming, computer controlled NC machines (CNC), and then the control of a number of CNC machines by a central computer (Distributed NC or DNC) as confidence in the technology increases.

As indicated, the next hurdle is to integrate these islands of automation - a hurdle that for most will take many years to achieve. Aerospace and automotive industries are well on their way and will pioneer many solutions for others to use.

You may recall the earlier reference to CAD tools being the firsthigh impact tools of the information age. The common denominator making these systems productive is information. In fact, information is the life blood of these systems - and it must be accurate, timely and of high quality. Simply stated, computers are tools that require high quality properly formatted information to function.

Quite frankly, few companies have existing manual (or otherwise) information systems that measure up. Many manufacturing companies exist today because people make them work. If you try to follow the flow of parts through a plant based on the documented information available, invariably one ends up talking to individuals who really know what is going on as little documentation actually exists.

Hence, when investigating the acquisition of a new computer based tool for a manufacturing application, an intimate and detailed look at the existing information system is essential. Only by so doing can a firm expect to achieve the productivity benefits touted by the salesman. Changes may well have to be made to the existing information system in order to obtain the full benefits of CAD/CAM technology.

A medium sized automotive parts manufacturer made the decision to invest in CAD/CAM technology based on the need to reduce time lost expediting parts and errors in the shop which had resulted in high rework costs. These costs had been with them for years but now they were having difficulty staying competitive.

In analyzing their needs it was shown that their information system had to be revised in order to enable the CAD/CAM system to achieve what they wanted which included producing an accurate bill of materials. Following the 16 month redesign and implementation period, the inventory and in-plant costs had dropped by some 60% and for the first time ever they had a handle on their work in process. The scrap and rework department was scaled down as the quality improved - all of this before acquiring the CAD/CAM system. The company was not only able to pay for the improvements from savings, but was able to defer the purchase of the CAD/CAM system by 2-3 years. Fortunately, they did not wait. After seeing the initial benefits, they forged ahead to gain the best possible competitive advantage.

The moral? Don't be mislead by promises that CAD tools are so "user friendly" that they enable you to do what you do the way you always did it. While this may be true - eventually - it may cost a considerable amount of time and resources to make the necessary adjustments to your present operation in order to gain the promised benefits. And that should not be a surprise. Can you ever remember buying a new tool that did not require some "getting acquainted time" on your part?

WHY CONSIDER CAD?

There are two main reasons for acquiring CAD today - the first arises out of sheer economic necessity (which is the primary driver today) and the second, arises from the desire of innovative and progressive people and firms to be on the leading edge to gain a competitive advantage in their market.

The latter are the entrepreneurs, the movers and shakers, the pioneers who will verify that the technology does or does not work.

To stay competitive today, quality in many cases now wins out over price. CAD technology provides both unmatched quality and low per unit part costs - if indeed, the proper hardware, software and humanware considerations have been included in the selection and implementation strategy.

A Case of Involvement:

Roy B. managed a small 43 man job shop in a Northeastern City. It seemed his shop was always so busy that they never had the time to investigate CAD/CAM. However, they received a directive that changed their operation forever. They were told that only those firms that could interpret digital data would be allowed to quote in the near future since only tapes and not drawings would be issued to quote from. Roy took it seriously. He knew his people lacked knowledge. But he also knew that some had shown high interest in CAD.

He knew too that with the current work load he could not afford to put many of his people on training programs. To save time he talked to a CAD/CAM consultant who carried out a quick survey of Roy's operation and suggested that to begin with, he should look at CAD to handle his small amount of drafting and some of his NC programming until the shop felt comfortable with the new tool. The consulting firm told him in no uncertain terms that he could not introduce new technology "any faster than your people can digest it".

Roy then talked to two of his most experienced young tool and die makers who had already been making noises about CAD. He asked them to look into what software packages might be applicable for their business and to select the appropriate microcomputer system they should purchase. He introduced them to the consulting firm and suggested they might save time by using the firm's experience to filter the hundreds of packages that appeared to be available.

The result: Giving the task to two tool and die makers who were already excited about the technology paid dividends. Both were excited. They spent many hours of their own time making sure the features offered actually did what the firm need. As Roy expected, once the system was in place they made sure it delivered.

They now have 3 IBM AT systems - one of which is booked every day of the week by new users wanting 'time on the system'. He is now looking at a larger system and is looking forward to making his next step with confidence.

The Case Of Methodical Gary:

Another 45-employee job shop firm in a north central community approached the task differently. The president and CEO was behind the investigation of CAD from the very beginning. He knew he had people on staff who were familiar with computers and their traits. He was sure that microcomputers just did not have the muscle for the needed calculation-intensive application. He dismissed them as not being needed as a tool to bring his staff up to speed. He too consulted a CAD/CAM consulting firm but only to "pick their brain" in order to gain an outside perspective as to where he might start, and also to see if he was missing anything. Gary's firm designed and built complex timing screws for the food processing industry and had enough of a backlog to enable him to spend time on a careful selection.

Looking back, they took approximately two years to select and install the right system for them. In the 16 months prior to the final decision and implementation phase, they reviewed the consultants report in detail, visited existing installations in North America and Europe, and attended trade shows with a long list of very focused practical questions. To their dismay they were shocked at the number of uninformed vendor staff representatives they talked to at the trade shows. They believe that many firms who do not do their homework would be at considerable risk in listening to "technology experts" who don't understand the application.

They did learn one very valuable lesson from the initial consulting company though – and that was to ignore terminology and features when talking to vendors... "talk only about what you need the system to do" they advise. In other words – know the what before the how (Chapter 3). This strategy saved hundreds of hours by enabling them to quickly detect those vendor representatives that could help them – and those who were wasting their time discussing system features that bore little relationship to what they needed.

The result: Gary's firm bought a $300,000 system that would have intimidated many firms his size and larger. But they bought the system they needed and it paid off. Today they are expanding it as new ways of using it are found. They are doing more work than they imagined and are hiring more people. They report their quality is the highest in the industry, their delivery times for new designs are the lowest, and they are now picking up unexpected export business.

TYPICAL SYSTEMS

The market abounds with technological overchoice - in other words we have so many choices that a road map is needed. There are now several hundred vendors providing advanced manufacturing technology and related products and services with over 50 claiming to be CAD/CAM system vendors. In 1986 the market grew to more than 4 billion dollars with the projection of a 22-25% annual growth to continue to the end of the decade. In other words, many companies - small medium and large - will be investing in new-to-them computer based tools. The trend is real. With the volume of all of this activity, it is no wonder non-computer professionals get confused.

1985 CAD/CAM ENGINEERING SEGMENTS
Total revenues $3.6 billion

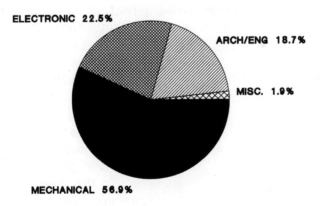

ELECTRONIC 22.5%

ARCH/ENG 18.7%

MISC. 1.9%

MECHANICAL 56.9%

Figure 4:   Note the three major CAD/CAM market sectors and their relative sizes in the mid 1980's

A good rule of thumb today is simply this.

First know what you want the technology to do - then select the software that will serve your need and only then look at the hardware required to run it.

To make life easier we can class available CAD/CAM systems according to:
1. The number of geometric dimensions they can represent
2. The manner in which they model, or represent, geometry
3. Their physical configuration

Geometric Representation:
There are basically three "flavors" - 2 dimensional, 2 1/2 dimensions and 3 dimensional systems.

2D systems: They are simply electronic drafting boards. Many high quality 2D software packages are available today running on personal computers(PCs). In fact, if you have a limited drafting need, or if you want to get started using CAD with minimum risk, PC based systems offer some outstanding capabilities in the $8,000-15,000 range fully configured. These prices are continuing to drop.

One drawback of 2D systems occurs when changes are needed. Changes must be made to each view in a multi-view drawing which is time consuming.

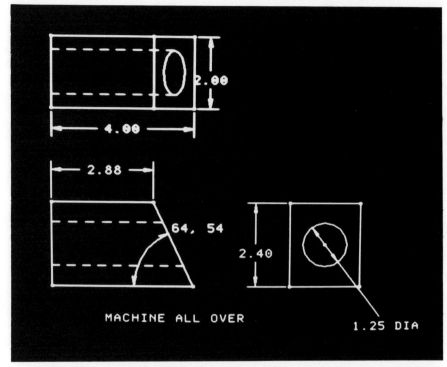

Figure 5.   A simple 2D representation is still sufficient for many small
applications.

2 1/2D systems: These have software that enables them to develop an isometric or perspective drawing from the 2D data provided.   This aids in interpretation but the drawing cannot be manipulated. Such systems may be based on PC's or on large mainframes costing hundreds of thousands of dollars. Professional Personal Computer based versions may run $4,000-8,000 for the software alone with mainframe systems costing many times more.   But these are dropping as well.

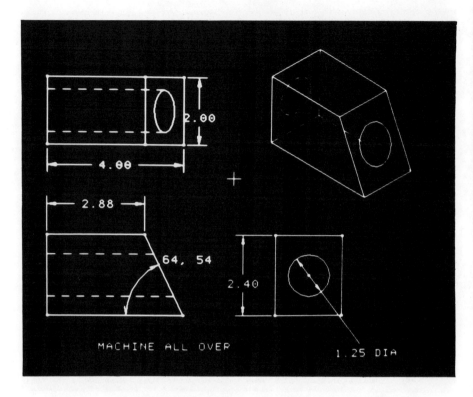

Figure 6.   2 1/2D representation aids interpretation of more complex
             parts and drawings

For those that need 3D systems, these provide a full 3-D
model of the part or component. Such systems allow the
designer to rotate the part and investigate it from any
perspective – a very definite advantage for most. By
changing any part of the geometry, all related views are
changed automatically. 3D systems are no longer simple
drafting tools – they produce full 3D models from which
quality drawings may be produced as needed. In fact
drawings are really now incidental. Such models offer the
ability to calculate centroids, moments of inertia, center of
gravity, volumes, and so on.

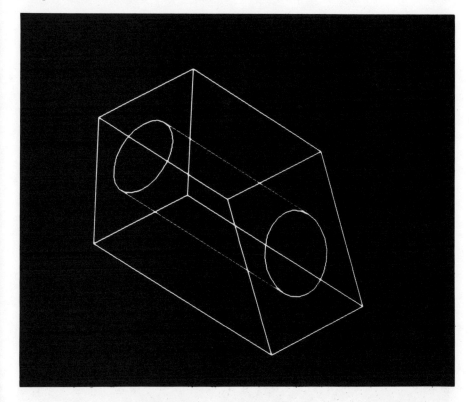

Figure 7.   3D representations are the most easy to visualize and interpret for most. They are the nearest representation to real models

Modeling Methods:
There are three major ones.

Wire Frame: With this technique, edges are identified by lines as the name indicates. While this is adequate for a great many applications, interpretation problems occur as part complexity increases resulting in an inordinate number of lines on the screen. These systems are fast since only line representations are manipulated by the computer. Since no point or surface definition exists off of the line, such systems have limited use in developing tool paths should one wish to produce NC programs.

However, by covering the surface of a wire frame model with a 'wire grid' that is very close together, cutter location files may be developed to drive NC systems.

Figure 8.   In this case lines define the geometric shape. There is no definition of the object in the computer that is not on the lines.

Surface: In this case surfaces are defined but any point not on the surface has no meaning as far as the computer is concerned. These are useful tools if modeling is needed and for the generation of NC tool paths. You cannot tell by looking at Figure 9 if it is a surface representation or a solids model of the part shown. Surface modelers provide much the same visualization capabilities but cost less than a solids modeler since the computer has to manipulate far less data.

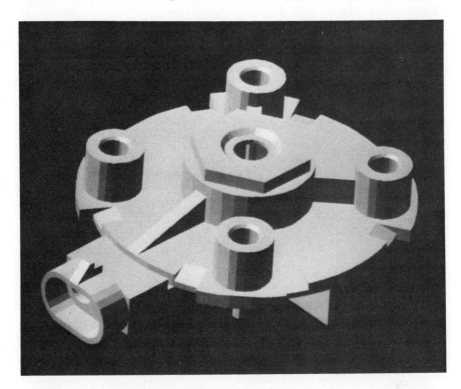

Figure 9.   A full solid model representation of an object defines all aspects of the object. Because all points on the surface and inside the object are defined in the computer, substantial computing power is needed.

Solids: In this case every point inside as well as on the surface of the solid has a location in the computer which means any reorientation of the modeled solid requires large computing capacity to recalculate the new locations of all the points.

Because of the need for such computing power, few solid modelers are available on PC's and those that do are slow and handle only a limited range of models. It is widely accepted that solids modeling software packages are the coming standard.   However, much work is under way by software developers to bridge the solid model to the machines needed to manufacture the part. As computing power and capability increases so too will the availability of solids modelers on PC's and larger systems.

System Configurations:
Again, three common configurations.

TODAY'S TREND

STAND-ALONE SYSTEM   **LOCAL AREA NETWORK**   **MAINFRAME COMMUNICATIONS (IBM,VAX ...)**

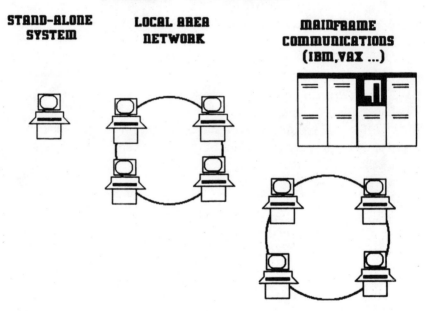

Figure 10.   Three basic configurations in widespread use today.

Stand Alone Workstation: As the name describes, the stand alone workstation is dedicated to a single user and consists of a means of communicating with the computer (keyboard, mouse, track ball, joy stick, etc), the computer, and an output device such as a monitor, printer or plotter. What you see is what there is. In the mid-1980's stand alone workstations on PC's ranged from a little under $10,000 to large minicomputer based systems running from $40,000-$150,000.

Local Area Network (LAN): As the name implies, a LAN ties together a number of microcomputers in order to share data bases, files, etc. There may be a few or many hundred computers in such a network. Depending on the type of system, costs can range from under $10,000 to $75,000 per seat. But again, such costs are dropping annually.

Host Based System: The most powerful of the configurations is the host -based system which relies on a mini or mainframe computer. Such systems can have almost any number of terminals and can provide a wide range of diverse software packages. Costs run from a few hundred thousand dollars and up.

Help is invariably needed by firms who initially do not have internal computer resources. Again, by focusing on what you want the system to do, many oversights can be eliminated. In this case it is wise to deal with either an "integrator" (a vendor who selects and combines hardware and software) or a "turnkey" vendor (one who makes the hardware and authors the software).

Although turnkey vendors can offer a degree of security when purchasing a system, the current trend appears to favor the integrators who combine the best hardware and software into a configuration that meets a given need and budget. As more software becomes available from third party authors the latter trend will no doubt continue. However, as your staff becomes more knowledgeable, they will play an increasing role in determining how your system will evolve.

PUTTING THE TECHNOLOGY TO WORK... THE STEPS TO PRODUCTIVITY

For strategies to work, an overall mind-set or philosophy must prevail as the pieces are assembled. The principle of setting objectives discussed on chapter 3 provides the framework for a workable implementation strategy. In addition, applying the CIM concept helps to direct the planning process.

If you are not a "computer person" the following rules will help shape your thinking and ensure that your venture into new technology will be a successful one with minimum risk. Post these rules over your desk as the business and technology action plans are being developed. And don't forget to include them as you go through the design iteration process.

These common sense rules can be golden:

Rule #1:   All new tools bring change. Be prepared to adapt to take advantage of the benefits they offer.

Rule #2:   Always determine the "what" before the "how". Be obsessed with determining exactly what the system must do before investigating the "how". Let no one deter you.(Page 49)

Rule #3:   The more accurately you know your needs the cheaper will be the system that will satisfy those needs.

Rule #4:   The benefits gained are geometrically proportional to your understanding of the information system that drives your plant.

Rule #5:   In manufacturing, parts cannot travel faster than the information that precedes them.
Hint:... Have you examined your information system recently?

Rule #6:   The magnitude and number of your problems will be inversely proportional to the amount of initial planning.

Rule #7:   Computers tools need quality and timely information. Document and quantify. Quality information is the life blood of advanced manufacturing technology.

Rule #8:   Garbage in is still Garbage out - a computer just gives it more apparent respectability.

Rule #9:   People make new technology productive. Plan with your people in mind.   Include them; listen to them.

Rule #10:   Technology cannot be introduced faster than your people can digest it.

Rule #11:   Plan as far as your horizon will allow - but move in the small steps you can afford and that your people can achieve.

Rule #12:   Plans and strategies that work are dynamic. Be prepared to iterate, iterate, iterate. (Chapter 1)

Rule #13   Purchasing CAD/CAM adds capacity.   Are you prepared for success? Have you a solid marketing plan that will enable you to feed the "hungry monster" with new orders once it's operational?

Rule #14:   Not all of the above rules apply all the time.

13 STEPS TO SUCCESSFUL INTRODUCTION

The Steps:
The following "nuts and bolts" steps were distilled from observing and consulting with hundreds of firms of which 1/3 were small, 1/3 medium, and 1/3 large. It is interesting to note that the steps are "generic" and apply regardless of the size of the company. Other studies have shown that of those that failed to reach expected results, invariably one or more of the following steps were overlooked or ignored.

Message: Take a hard look at the steps and ensure your approach incorporates them.

Step One: Understand the Technology. Learn what the technology can and cannot do. This can be done in less time than you think.   Visits to the plants of current users, attendance at trade shows, and introductory seminars provide a basic understanding of the capabilities of the technology. Every professional must set aside time to keep up to date — it is more than a professional obligation - it is simple economic survival.

For a more in-depth understanding, investigate what programs and courses are offered by area Colleges or Universities.   Contact the Society of Manufacturing Engineers, the Institute of Electrical and Electronic Engineers or other professional organizations to see what they have to offer or would suggest.

Step Two: Define The What... Before The How. You will remember this from Chapter 3 "Defining the Project Objectives" and should commit this phrase to memory. Remember it each time an enthusiastic proponent of the technology begins to seduce you with its features.

This is the pivotal step since all steps to follow depend on its accuracy.   Simply apply the common sense approach of taking a hard-nosed assessment of exactly what it is you want to do. Don't be mislead into discussing the "how" to early — that will come later.

Keep in mind that the more accurately one can determine their needs the cheaper will be the system that will satisfy those needs(Rule #4).

Step Three: What Type Of System? The better you are able to identify your needs, the numbers of vendors offering appropriate products melt. In addition, the type of system that will best suit your need begins to emerge automatically. It is not unusual to end up with only 3-4 vendors after completing a rigorous needs analysis. Never hesitate to visit existing installations to learn from others - in many cases this is the most objective information you can obtain..

Step Four: Can You Cost Justify It? Now that you know what you want to do, and you suspect which type of system is needed, a simple no-nonsense cost justification should be drafted. Keep in mind that there will always be companies that cannot use CAD/CAM or robotics profitably — and yours may be one. Be sure to take into account those hard to quantify measures such as reduced errors and quality. But even figures for quality may be arrived at indirectly by examining current scrap rates and rework costs.

Forcing yourself to put numbers on such benefits raises solid questions that might not be raised otherwise. Further, why not learn from the experience of others by considering consultants who have helped similar firms justify their systems.

Step Five: System Specifications. You should now be very clear about what the system must do and which system configuration will probably do the job. Now is the time to turn to the vendors to confirm what is really available. This strategy yields two big benefits. The first: since you know what you want, you know what questions to ask and can keep control of the discussion while forcing the vendor to focus on your application. In short, you won't be mesmerized by unnecessary features.

The second major benefit accrues to both of you. Professional vendors sincerely appreciate talking to prepared and knowledgeable individuals who know what they need. The time saved will be substantial and will be of mutual benefit.

A hard look at the near-future capabilities will indicate what flexibility the new system should have. However, a number of trade offs will have to be made. For example, how realistic are the desired features – will they put the cost out of sight? As a result, a system can be specified with growing confidence on your part and reviewed with vendors to determine what products will work. They often have excellent suggestions to make.

Step Six: Shortening The Vendor List. Completion of steps 1-5 automatically shorten the vendor list. Look closely at things like "vendor satisfaction" and don't hesitate to ask vendors for the names of firms with similar installations. And call them. They can only say no – and they normally don't. What you can learn from their experience can be invaluable.

Reduce the list of vendors to approximately three. Any more can cost you a lot of time verifying results, traveling, etc, etc. Now you are ready to benchmark the systems and see if they really can do what they claim... before you are firmly on the hook.

Step Seven: Benchmarking. This is your last chance to confirm that the system will do the job before you are committed.

Benchmarks cost money and you may encounter some vendor resistance. However, by being prepared and by providing clear guidelines of what you are looking for you should obtain excellent cooperation.

A benchmark is simply a comparative performance test (or task) that will be run on a machine configured exactly as the one you are proposing to acquire. The benchmark or task should not be given to the vendor ahead of time. However, to be fair, a clear outline of what you want to do must be communicated in order to ensure the right equipment and a competent operator are present. Choose a typical routine project/task that you want the new system to do for you (one that will have a high payoff). It should include all the functions you need but not the exceptions you encounter only rarely – again, be fair. The benchmark should then be taken to the vendor's location and executed and the results tabulated.

Step Eight: Software/Hardware Selection. Bear in mind that no matter what system you purchase, there will be problems. That's life. That's reality. In many cases, more than one vendor's system will do the job – so how do you make the final decision? Making any new system productive requires good old fashioned "hand-holding". Ask yourself... which vendor will provide the after-sales support? What vendor would your people best like to work with? Which vendor do you trust most? Will the vendor be in business over the next 5 years to support you?

And, don't overlook your people in the process. Which system do they want? Keep Rule #8 in mind because it will be your people that will make the system work. By purchasing a system that they want, they will do their utmost to make their system work.

Step Nine: Plans and Procedures. Now that the system is defined, benchmarked and selected, the procedures needed to make it functional must be finalized. Who will maintain it? When and how will backups be carried out? Who will do the back ups? Will the pay scales change? Who will receive priority access to the system? Who will be trained? Will there be 2 shifts or 3 shifts – and how will they be organized?

How will the data base be organized? Who will ensure that the engineers don't mess up the files? So many questions to be decided that a the "should I talk to a consultant or not?" question should be raised. Fortunately, since many firms have gone through this exercise already, their experience is readily available – for a price. But be careful. Seek advice from a firm or a consultant that has related experience.

Step Ten: Training. Most companies underfund training. As we move into the information society the need for continuous training will become more obvious. Two key factors cited in the failure of systems to not meet their objectives were:

1. Lack of adequate planning
2. Lack of adequate training – both initial and ongoing

When budgeting for training, ongoing training cost estimates may vary from 1% per month of the installed equipment costs for a CAD/CAM system to a gross figure of 150% of the hardware/software costs for a large system with CIM capabilities that requires much retraining.The important factor to note is simply that the duration and cost varies with the complexity of the application. On going costs should always be included. It is human nature to become satisfied with a given way to function. Outside training will keep your staff alert to better ways of applying the expensive equipment you have acquired.

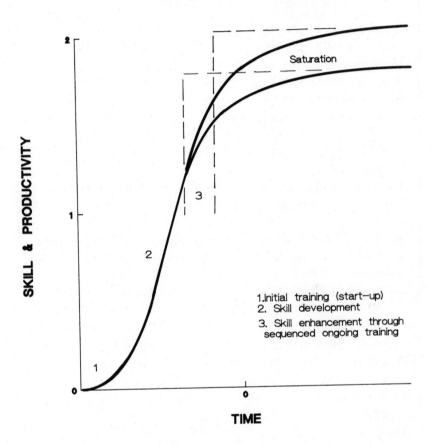

Figure 11.   The generic 'learning curve'. The horizontal scale
depends upon the application for which an individual is
being trained and its inherent complexity.

Figure 11 shows the standard learning curve. Although the shape is quite generic, the time scale will vary depending on the application.

A small electronics component manufacturer installed a PC based CADD system. The firm produced a large quantity of ladder diagrams and basic schematics that incorporated the same components over and over. They approached training in this way. Three staff members attended a 2-day hands-on training program away from the plant. This was followed by a week of in-plant training where all of the standard components were entered into the CADD system.

Jack refers to his CADD system as his "never-having-to-draw-anything-twice-system" (sound familiar?) because schematics are drawn in 1/6 to 1/10th of the time because the symbols are all in memory. Jack estimates that to really get people up to speed for his application requires only about 3- 6 months.

The learning period will vary. If you are training a mechanical designer to use the full capabilities of a large CAD system which may include finite element analysis, kinematics, and other advanced capabilities, it may well take 12-18 months for the individual to apply the system fully. This program will require daily system work plus periodic training sessions on the vendor's premises. This approach helps guard against the tendency for an operator to learn just a few functions and not the rest.

A medium size firm headquartered in Ohio recognized the competitive advantage CAD, CAD/CAM and CAE tools could have in designing and manufacturing electromechanical systems and printed circuit boards. They were also aware that fear of job loss existed if high capacity systems were installed. To ease the potential morale problem the firm announced to all employees why it was critical for their survival to employ the new technology. It was clearly stated by senior managers that the system was justified on the basis of quality improvements, reduced scrap and rework, and shorter design cycle times – not on staff reductions. No staff reductions were considered. Further, training programs were to begin immediately for all staff – those both directly and indirectly involved – in order for everyone to learn together.

Three training programs were set up immediately. One for senior managers who needed to understand how this new technology would impact their span of control, one for those who would be responsible for system productivity (department managers) and one for the operators. In addition, system time could be booked by anyone in the plant.

The result: A smooth transition. A planned strategy based on open communication and respect for employee concerns and contributions pays dividends. In fact the individual in charge of the implementation strategy reports "we are doing far more with the systems than originally planned". "Staff continually come up with new ways of applying the system. I don't think we would be getting this kind of contribution if everyone had not been involved from the beginning". He added "Some believe it is their system – and that's just great for us".

Normally, CAD operators are drawn from existing employees. CAD systems are set up to be compatible with standard drafting practice which is a help when introducing the technology to older users. Roughly 2-3% of employees who are converted to CAD find difficulty so it is reported. On the contrary, reports abound that operators feel less tired at the end of the day, have far fewer errors, find more time to check designs and take more pride in their work.

The big benefit to the company lies in the development and utilization of the "data base" which older employees previously carried around in their heads. Putting CAD tools into their hands produces quality designs drawn from company-specific experience.

Remember, it is the "humanware" that will make the new tools productive.   Training costs are investments, not expenses. And an investment in people will yield multi-fold benefits. Do not underestimate the costs – in fact be liberal but again, check with someone who has trained staff successfully and hear their story.

Step Eleven: System Installation. Installation considerations must include such items as:

* Consideration for loss of productivity due to disruption
* Construction of the environment which may include a temperature controlled computer room, changes in lighting, new chairs/furniture and provision for enough elbowroom to lay out drawings etc.
* Human factors considerations. Keep in mind that operators may be sitting in front of terminals for 6-8 hours per day. Seating, lighting, and office layout considerations take on added importance and advice from a specialist can avoid morale as well as eyesight and back problems. When video display terminals were first introduced, the vast majority of the

complaints arose from lighting and seating problems. People just were not used to sitting in front of a video screen for hours at a time. Extra care here will reap dividends.

The installation will represent change to people... and change represents uncertainty and uncertainty breeds fear. Ensure all involved are fully informed in advance of what will happen and how it will affect them. If you have done it properly, the system's implementation will just be another event on the planning schedule.

> George W. was not sure how his employees would react to a CAD system so he used a summer student in her senior year to introduce CADD into his 23 man shop. Since she was part time she did not threaten anyone and they were curious to see what she was up to as she began building a data base of standard components and products. With George's help she set up informal evening sessions on the computer for anyone who wished. Those who wished could book time on the systems from 6:00 pm to midnight. These activities turned out to be so well received that George acquired a second PC.

Step Twelve: Measuring Results. Almost every system installed has soon shown a need for expansion. To expand costs money which in most cases requires a new proposal and request for acquisition of funds. You can be sure that one of the first questions your financial people will ask is "what are the results of the first installation? Has it justified the expense?" In other words, whether you get the money or not may well depend on how well you complete the initial steps — and how well you measured your progress.

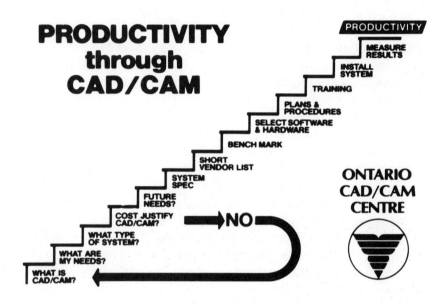

Figure 12:   The "Steps To Productivity" shown here were developed
              by the Ontario CAD/CAM Centre based on the
              compilation of contracts with 500 companies of all sizes.

Step Thirteen: Expansion Considerations. The first questions
to be faced will no doubt be "Can we build on to our present
system?" Or, the second may be "Can we discard the first
system and acquire the next largest size?" or "Can we do
both?" A return to one's basic objectives will provide the
appropriate answers.

PROBLEMS, TEMPTATIONS AND PITFALLS

The first pitfall to avoid is the subtle and very pleasant seduction of the technology. A potential user visiting a trade show who has not followed the above steps will assuredly gain a host unrealistic expectations. These unrealistic expectations are blamed for many systems not meeting spec.

Fight that temptation to get excited about the system features... because many of them you will never need. Ruthlessly apply rule #2 and keep forcing the discussion to focus on what you need the system to do.

Be careful of the "I'm going to do it all myself" trap – or the "nobody knows my business but me" refrain - or the one that that goes.. "my people can do most of this on their own time - we just can't assign anyone to it full time right now" Lets face it, the rate of change in technology in the last 5-10 years has made it almost impossible for anyone to keep on top of it let alone stay competitive in the market place and survive.

Make no mistake though, there really is no one else that knows your business better than you – but there are many that know the technology. So once you know what you really want to do, the technology experts can help you and probably reduce your implementation time by 6-18 months. Its simply called "learning from the experiences of others", or, to put it another way it's "re-invention-of-the-wheel-avoidance".

To select the right consultant, pick one that works in your field, understands your problems, has a proven track record – one who will readily share the names of former clients with you. When you find one don't forget to call their former clients.

People Considerations: People will make the technology productive, or, will ensure it never reaches the figures targeted. So don't skimp on training or in communicating with them.

HOW DO I KEEP UP AND – WHAT'S IN IT FOR ME ANYWAY?

To answer this question, take a hard-nosed perspective. What are your personal and professional objectives? What does CAD really have to offer you? Probably two major benefits will affect your career path. The first is professional development. Firms are advertising for professionals with a CAD background – and those that are not are beginning to think about how CAD may be introduced into their companies. Candidates who have CAD exposure are getting extra consideration.

The second benefit is personal development. Quite simply your personal worth on the open market will be enhanced by a solid understanding of how to apply new computer based tools such as CAD/CAM. Your skills as a designer or project manager will remain marketable, however, the extra leverage CAD skills can provide enables you to demand a higher salary while at the same time providing you with much increased job satisfaction.

As a practicing professional, you will be building on an existing experiential knowledge base which a young graduate cannot have and that can be worth money. As a young graduate, the chances are very high that because of the visualizations capabilities and quick reference to alternative designs offered by CAD tools, you will be able to reach a practicing level of proficiency more quickly on the job.

I believe many chief executive officers want to head progressive and innovative companies which apply leading edge technology. There is much prestige and self satisfaction in building a winning team that is capable of carving out a new or increased market share. The "tools" they will need are really people who can apply the technology.

Designers and Engineers who can help a firm introduce new technology by bypassing problems and providing solutions are – and will continue to be – in high demand. Enough said.

ASSIGNMENTS:

1. Write down, in order of importance to you, three learning objectives that you want to achieve which relate to CAD, ADD, CAD/CAM or CIM technologies.

2. With the three objectives in mind visit your plant, school,or local public library. Spend a full hour reviewing the engineering and technology periodical section. Extract and copy one article that addresses each of your three objectives.

3. Review the card catalogue and look through the CAD/CAM texts and reference materials. List on a 3"X 5" card the names of three reference sources that you think may be useful to you.

4. Ask your instructor for the page numbers in this chapter of 3 Case Histories. For each one, indicate what they did well or poorly according to the recommended steps in this chapter.

5. What does CADD do for the design process? Refer to Chapter 1 and develop a 1-paragraph statement.

6. List at least three benefits that may be obtained by a manufacturer who successfully introduces a CAD/CAM system. List two potential problems that could delay or spoil the successful introduction.

Appendix I
How to Use This Book
with Courses on
Engineering Design

Prof. G. Kardos, BSc, ME, PhD
Professor of Engineering
Carleton University, Ottawa, Canada

Prof. Kardos is the author of 15 case studies for the ENGINEERING CASE LIBRARY.

EXCELLENCE THROUGH PRACTICE

In this book Syd Love has given us a guide to mastering the difficult task of doing Engineering Design. The application of these principles can turn the activity from a seemingly random effort—dependent on serendipity for success—into a systematic, dynamic and efficient means of achieving a well defined end. But the operative word here is *application*, simply knowing the principles is not enough. To achieve *excellence* in design, one must practice these concepts in a creative and skilful manner. Only when the practice becomes instinctive can its full potential be realized.

It must be apparent to any discerning reader that the principles outlined by Mr. Love cannot be used as a simple recipe to be followed mechanically. The designer must carefully assess the situation, decide where he is in the design process and select the techniques and methods that will best assist him achieve his objectives. *In short, the techniques must be applied with judgement.*

How can one develop the required experience? How can one receive the practice in applying these principles? Especially in the context of the real design situation, where the issues are never clearly delineated and conflicting requirements and personalities often seem to contradict one another.

To the practising engineer the answer is obvious, these concepts sould be implemented on the job immediately. At first their use will be strange, but with practice they will come easier and soon pay huge dividends.

But how does the student engineer practice these principles? He does not have the opportunity to practive them on a job. Must he learn them only in theory? Must he wait until he can learn through experience? Not necessarily! There are ways that a student can practice the use of the ideas presented in this book in a context that simulates real life.

DEVELOP "PROS" THROUGH DESIGN PROJECTS AND ENGINEERING CASES

There are two useful ways for a student to learn about design and both involve practice. The first is through design projects, the closer these projects go toward producing hardware the better. The second is through the use of Engineering Cases. I have used both modes of instruction in a number of schools and at various levels and have found that both modes can advance the engineering student toward becoming a competent professional.

The use of design *projects* gives the student an opportunity to experience for himself *the thrill of doing design* and provides the satisfaction and the confidence that comes from accomplishment. If the project is well formulated, the student will be able to accomplish a great deal in the allotted time. But classroom projects cannot show the many faces of engineering reality. Because no single project will ever encompass all the possible difficulties that may be encountered and the classroom cannot reflect the kind of commercial pressures that so often determine the course of real-world design. However, in spite of this, *students should be required to do design projects* to demonstrate to themselves that the various topics they have learned in engineering courses can be brought together, used for synthesis and result in a desirable and practical object. The chapters of this book are an excellent guide for students doing design projects.

However, to learn and practice the principles outlined in this book, I think that the far more important medium is the *Engineering Case*. Here the student has *a slice of real engineering* that he can analyse, that he can practice on.

An "Engineering Case" is a documented record of an engineering activity as it was actually carried out. It is not a description of a project as "it should have been carried out," which we usually get in technical reports. An Engineering Case presents what actually happened as closely as possible, failures as well as successes. Since any documentation must be subjective, the case is usually presented from the point of view of one of the participants. The case contains not only the technical details; it is also fleshed out with the financial, time, facilities and other constraints as perceived by the principal in the case.

The "Engineering Case" is the engineer's equivalent to the "Business Case" well established in the Business Schools. The parallel is not coincidental. Cases are used in Business Schools to teach their students to make decisions in difficult, complex, and fluid situations. Syd Love's book gives us guidelines for managing the design process: what better way to learn these complex design decisions than through Engineering Cases.

In using Engineering Cases I have noted that the *students quickly identify with the people in the case* and tackle the problems with the same level of commitment that would be expected in practice. In fact, the greatest difficulty with a Case session I find is to end the session: to tell the students that we have accomplished what we had set out to do with the Case and it is time to move on to a new phase. Outside reality is brought into the classroom with Engineering Cases.

TWO APPROACHES TO THE USE
OF ENGINEERING CASES

Engineering Cases can be used in many ways depending on the style of the instructor and the objectives of the session. Cases can be used to reinforce the ideas presented by Love in two ways:

1. The student can critically examine an Engineering Case and evaluate what has happened in the light of what has been presented by Love.
2. Selected breaks in the Case can be used as a "critical instant" and the student is required to decide what to do next.

In the first approach the student has an opportunity to act as a "Monday morning quarterback" to *evaluate* the moves of the "pro," to consider the alternative actions that would have achieved better or more efficient results. This form of analysis sensitizes the student to situations where he can use the various techniques. His critical examination with respect to Love's principles will provide an opportunity to match a suitable technique with the right situation.

And *careful assessment of how the situation could have been improved* means that the lessons in this book will be reinforced. The student will become an advocate of these principles instead of a passive recipient.

With the "critical instant" approach the student can no longer function as a passive observer weighing the merits and demerits of a situation. He is plunged in as an active participant. He is required to *decide;* the onus is on him to put into practice what he has learned. He must select the appropriate action and implementation. *In dealing with these principles within the context of the real world situation, the subtleties of the technique are brought home to him.* In seeing how others react to his suggestions and proposals he will have an opportunity to test the merits of the methods and his understanding of them. He has *a chance to practice what he has learned.*

Of course, these two approaches need not be treated separately. Which is emphasized will depend on the Case being considered, the student's reactions and how far into the course the class has progressed. Many times the class situation cries out for a shift from passive observer to active participant, such as when the class's analysis becomes supercritical, the realities of the situation can be brought home by asking "What would you do?". The more specific the recommendation asked for, the more constructive will be the response. Similarly, when the students are having difficulty in coming to grips with the situation, it usually means they don't understand it properly and are plunging in without proper preparation. By asking "What's going on here?" the direction is changed toward increasing understanding.

MODES OF USE

Engineering Cases are not only used with different points of view, but they can be used in different modes. It must be emphasized that there is no right or wrong mode; all modes of using Cases will accomplish something and the instructor must select the mode that he and his class feel most comfortable with, and which accomplishes his objectives. The only guiding principle that I would suggest is that the learning is maximized the closer the student identifies with the case. The amount of effort he puts in is a measure of what he will learn. I like *two modes* of Case use best:

1. *The directed class discussion,*
2. *The written assignment.*

For *class discussion* the students are assigned the case for reading before the discussion period. In the early stages, the questions to be considered are also assigned; in the later stages this is not necessary or desirable. The class period is used as an open discussion in which the issues in the Case are discussed by the students, starting with the assigned questions but not restricted to them. *The instructor takes part in the discussion as little as possible.* He monitors and guides it, but he expects and insists that the students do the thinking and decision making. He may at the end suggest further considerations and he may on the basis of his evaluations of how well the students have mastered the objectives of the session make reading assignments or assign activities to be completed for the next session. Such discussions can be stimulating and give the student the feeling and confidence that he has discovered for himself the vital conclusions to be drawn from the Cases.

The written assignments can be long or short. I prefer longer assignments when I want to *cover the sweep of a whole project,* such as analysing the various stages or in the design processes. Shorter assignments usually take the form of role playing in which the student is asked to take the part of the engineer in the case or some peripheral character and to make a *written recommendation* to the firm involved.

Such written recommendations take on a different character than oral responses in class. They are more thoughtful and the issues are more carefully considered. The oral discussions allow for a great deal of free wheeling; the written presentations bring these free wheeling ideas down to earth. Of course, *written assignments can be required after a free wheeling discussion so that the student can consolidate and rationalize the many ideas he had to deal with in the discussion.*

ENGINEERING CASES THAT YOU CAN USE

In reading this book you have been exposed to several short cases that Syd Love has used to illustrate his points. But because they have been tailored to the needs of the text they cannot always be used as classroom cases. Fortunately we do not have to depend on our own resources to find suitable cases. *The American Society of Engineering Education* has a collection of over *250 cases* which they are willing to share with you at the cost of reproduction.[1] I recommend them highly.

A quick selection of cases by discipline is shown in Table 1.

For a classification of 200 cases by 8 disciplines see "Engineering Cases," a reference at the end of this appendix.

The following is a synopsis of some typical cases and how they could be used in conjunction with the chapters in this book.

ECL 1–13 *"Development of an Oil Well Stripper Rubber."* This is a two part case. The first part is the formal report on the engineering. The second part is a candid narrative recalling the various adventures within the firm and in the field which were necessary for successful development. It is useful with chapters 1, 2, 3, 6, 10 and 12.

ECL – 63 *"Data International II, Voltage Regulation."* This case is in three parts. It deals with the search for an inexpensive means of reducing the voltage fluctuations of the voltage supply in a remote mission in Guatemala. The solutions of several consultants are given. They must be evaluated and matched to the requirements. It is useful with chapters 2, 6, 7, 8, 9 and 10.

ECL – 77 *"Request for a Bridge Design."* A 100 foot bridge is needed on a road leading to a 1,500 megawatt power station. Using handbook data the engineer chooses a standard design with tradeoffs between time, weight and cost. Hidden within the case is the unstated difficulty that the engineer is solving the wrong problem. It is useful with chapters 2, 3, 4, 8 and 9.

ECL – 114 *"Development of a Circular Strike Plate."* This describes the complete development of a new piece of door hardware. Changing technology demands a new form of strike plate. The case describes the various stages that the design had to go through to satisfy customer, patent infringements and manufacturing. The buy or make decisions are implicit within the case. It is useful with chapters 1, 2, 4, 5, 6, 7, 8, 9 and 13.

ECL – 129 *"To Support a Mountain."* Because of its unique properties wood has been used for propping mine tunnels. This case describes the development of an advanced form of hydraulic mine prop. It shows the difficulties that had to be overcome to perform in a manner similar to a wooden prop. The case covers two iterations of the complete design. The final section covers the design evolution of a single component—an unloading valve. It is useful with chapters 1, 2, 3, 7 and 9.

ECL – 135 *"Problem of the Perverse Pinion."* This is a case in failure analysis. An automotive starter pinion has a higher warranty replacement rate than anticipated. The material selection, design and production process are examined to determine cause of failure. Recommendations are made. Eventual follow-up reveals that the recommendations, although accepted, were not implemented;

financial consideration led to an alternate action which did not cure the problem, but which did reduce it to an acceptable level. It is useful with chapters 3, 5, 8, 9, 10 and 11.

ECL - 166 *"Moving Map Display."* The engineer in this case is required to produce an improved version of a device that automatically displays the plane's position. His task is complicated by the fact that he has minimal information about the original design and he knows he is competing with another firm which has a head start. The case details his design up to and including field testing the first prototype. The epilogue outlines the successful sale and production of the hardware after the engineer has left the company. It is useful with chapters 1, 2, 3, 4, 5, 7, 8, 9 and 12.

ECL - 167 *"Methods of Lessening the Consumption of Steam and Consequently of Fuel, in Fire Engines."* This case chronicles James Watt's pivotal invention of the steam condenser told in his own words and his contemporaries. The interaction of creativity, science and experimentation is illustrated. It is useful with chapters 2, 3, 6, 7 and 10.

ECL - 177 *"Fairdeal Construction Company."* This case shows how a construction company bid on a contract based on drawings and vague information supplied by the customer. Underground water conditions prove to be different than expected and action must be taken. It is useful with chapters 2, 3, 5, 9 and 12.

ECL - 188 *"The Stun Gun."* The turbulent sixties created a need for a nonlethal weapon to counteract civil disobedience. This case outlines how an engineer invented and developed a gun that fires an oversize bean bag to satisfy this need. It is useful with chapters 2, 3, 4, 6, 7, 8 and 10.

ECL - 206 *"Spectrac Limited."* This case describes the design and development of an "add-on" device that will convert black and white television sets to colour sets. The design is technically successful and highly praised. The difficulty arises in raising money because of the doubt about its marketability. It is useful with chapters 2, 3, 6, 7 and 11.

ECL - 208 *"The Snotruck."* This case details the conception, design, and development of a unique kind of snowmobile. It shows how a market survey was used to define the need. Details are given of technical difficulties that were overcome through three prototypes to produce a product that the developers felt was ready for the market. This case is useful with almost every chapter in the book.

ECL – 213 *"Paul Hait and the Dental Unit."* This case tells of a creative young engineer who is asked to take over technical direction of a company founded to exploit a dentist's invention. How he deals with the internal conflicting requirements is told in the first person. Commercial success leads to the collapse of the company. It is useful with chapters 2, 3, 5, 6, 7, 8, 11 and 12.

ECL – 214 *"Boom-Boat Drive and Steering System."* This case covers the development of a unique drive system. It shows the early ideas and prototypes that were necessary to make the drive work properly. Then the whole concept was thought through again in the light of what was achieved and a much simpler design evolved. It is useful with chapters 1, 2, 3, 6, 7, 8, 10 and 11.

ECL – 222 *"Measuring Motor Torque-Speed Curve."* Because the firm designs and builds electric motors a need exists for an instrument that will quickly and accurately display the speed-torque characteristics of a motor. The chief engineer conceives a schematic of a system and finds the money and an engineer to do the development. The case outlines how the engineer went about designing and optimizing the necessary circuits. The resulting instrument immediately provided insites into motor operation that would yield dividends far beyond the cost of the instrument. This case can be used successfully with all the chapters in the book.

ECL – 234 *"International 500."* This case describes the development of a new model of a manure spreader. The case shows how creative free thinking was used to produce the first prototype and grab the interest of marketing. Successive tradeoffs and field tests eventually evolve the design into a product that can assure an increase in the company's share of a very competitive market. This case can be used successfully with all chapters of the book.

These are only a few of the cases that can be used successfully to illustrate the principles that Love presents in this book. The cases demonstrate that these principles can be applied within any discipline. The cases show that the application of the principles leads to success. Where the principles are not properly applied their use could avoid some of the difficulties encountered.

I have suggested that the use of cases in the classroom situation is beneficial. The student without engineering experience needs real world examples within which to set these ideas. The practicing engineer can also benefit from studying these cases in the light of Love's principles. Using the cases instead of his own experience permits him to view the problems with a degree of detachment which he cannot have with respect to his own experience. Therefore,

I recommend the use of Engineering Cases to all those who intend to benefit from this book.

References:

1. *Engineering Cases—A List of Engineering Cases,*
   American Society for Engineering Education, One Dupont Circle, Washington, D.C. ($1)
2. "Outside Reality Inside the Classroom"—H. O. Fuchs,
   Engineering Education, March 1970.
3. "Engineering Cases—Feedback from Industry," G. Kardos,
   ASME Paper 74-WA/DE-31; 1974.
4. "Pointers on Using Engineering Cases in Class," G. Kardos;
   Engineering Education, January, 1978.
5. "An Easier Way to Teach with Engineering Cases," Karl H. Vesper;
   Engineering Education; January, 1978.

	Case	Iteration (1)	Needs (2)	Objectives (3)	Criteria (4)	Tradeoffs (5)	Creating (6)	Feasibility (7)	Selection (8)	Communication (9)	Optimization (10)	Time & Cost (11)	Management (12)
Civil	ECL–77		0	0	0				0	0			
	ECL–177		0	0		0					0		0
Electrical	ECL–63		0				0	0	0	0	0		
	ECL–206		0	0			0	0				0	
	ECL–222	0	0	0	0	0	0	0	0	0	0	0	0
	ECL–166	0	0	0	0	0		0	0	0			0
Mechanical	ECL 1–13	0	0	0			0				0		0
	ECL–114	0	0		0	0	0	0	0	0			0
	ECL–129	0	0	0				0		0			
	ECL–135			0		0			0	0	0	0	
	ECL–167		0	0			0	0			0		
	ECL–188		0	0	0		0	0	0		0		
	ECL–208	0	0	0	0	0	0	0	0	0	0	0	0
	ECL–213		0	0		0	0	0	0			0	0
	ECL–214	0	0	0			0	0	0		0	0	
	ECL–234	0	0	0	0	0	0	0	0	0	0	0	0

A Quick Selection Chart for Engineering Cases

Index